# REPLICATION IN EXPERIMENTAL ECONOMICS

# RESEARCH IN EXPERIMENTAL ECONOMICS

Series Editors: R. Mark Isaac and
Douglas A. Norton

Recent Volumes:

| | |
|---|---|
| Volume 7: | Emissions Permit Experiments, 1999 |
| Volume 8: | Research in Experimental Economics, 2001 |
| Volume 9: | Experiments Investigating Market Power, 2002 |
| Volume 10: | Field Experiments in Economics, 2005 |
| Volume 11: | Experiments Investigating Fundraising and Charitable Contributors, 2006 |
| Volume 12: | Risk Aversion in Experiments, 2008 |
| Volume 13: | Charity with Choice, 2010 |
| Volume 14: | Experiments on Energy, the Environment, and Sustainability, 2011 |
| Volume 15: | New Advances in Experimental Research on Corruption, 2012 |
| Volume 16: | Experiments in Financial Economics, 2013 |
| Volume 17: | Experiments in Macroeconomics, 2014 |

RESEARCH IN EXPERIMENTAL ECONOMICS VOLUME 18

# REPLICATION IN EXPERIMENTAL ECONOMICS


EDITED BY

**CARY A. DECK**

*University of Arkansas, Fayetteville, AR, USA*

**ENRIQUE FATAS**

*University of East Anglia, Norwich, UK*

**TANYA ROSENBLAT**

*University of Michigan, Ann Arbor, MI, USA*



Emerald

United Kingdom – North America – Japan
India – Malaysia – China

Emerald Group Publishing Limited
Howard House, Wagon Lane, Bingley BD16 1WA, UK

First edition 2015



**British Library Cataloguing in Publication Data**
A catalogue record for this book is available from the British Library

ISBN: 978-1-78560-351-8
ISSN: 0193-2306 (Series)

# CONTENTS

# LIST OF CONTRIBUTORS


| | |
|---|---|
| *Nomin Batmunkh* | Department of Economics, University of Alaska Anchorage, Anchorage, AK, USA |
| *René Bekkers* | Center for Philanthropic Studies, Faculty of Social Sciences, VU University Amsterdam, Amsterdam, The Netherlands |
| *José Gabriel Castillo G.* | Department of Economics, Texas A&M University, College Station, TX, USA |
| *James C. Cox* | Experimental Economics Center and Economics Department, Georgia State University, Atlanta, GA, USA |
| *Catherine C. Eckel* | Department of Economics, Texas A&M University, College Station, TX, USA |
| *Haley Harwell* | Department of Economics, Texas A&M University, College Station, TX, USA |
| *Alexander G. James* | Department of Economics and Public Policy, College of Business and Public Policy, University of Alaska, Anchorage, AK, USA |
| *Duncan James* | Economics Department, Fordham University, Bronx, NY, USA |
| *Xu Jiang* | Fuqua School of Business, Duke University, Durham, NC, USA |
| *John A. List* | Department of Economics, University of Chicago, Chicago, IL, USA |
| *Radhika Lunawat* | Merage School of Business, University of California – Irvine, Irvine, CA, USA |

| | |
|---|---|
| *Zacharias Maniadis* | Department of Economics, School of Social Sciences, University of Southampton, Southampton, UK |
| *James J. Murphy* | Department of Economics, University of Alaska Anchorage, Anchorage, AK, USA; Nankai University, Tianjin, China |
| *Benjamin Nilsson* | Department of Economics, University of Alaska Anchorage, Anchorage, AK, USA |
| *Samantha Ray* | Department of Economics, University of Alaska Anchorage, Anchorage, AK, USA |
| *Brian Shapiro* | Opus College of Business, University of St. Thomas, St. Paul, MN, USA |
| *Jason F. Shogren* | Department of Economics and Finance, College of Business, University of Wyoming, Laramie, WY, USA |
| *Fabio Tufano* | School of Economics, University of Nottingham, Nottingham, UK |
| *Adam Zylbersztejn* | Université de Lyon and Université de Lyon 2, Lyon, France; CNRS, GATE Lyon Saint-Etienne, France |

# ABOUT THE EDITORS

**Cary A. Deck** is a Professor of Economics in the Sam M. Walton College of Business at the University of Arkansas where he serves as the Director of the Behavioral Business Research Laboratory. Dr. Deck is also a Research Professor at the Economic Science Institute at Chapman University. He has published many scholarly articles in the field of experimental economics. His research has been funded by the National Institutes of Health, the National Science Foundation and other organizations.

**Enrique Fatas** is a full-time Professor of Economics at the University of East Anglia where he serves as the Head of the School of Economics. His interdisciplinary research has been published in *Economics* journals and other journals like the *Proceedings of the National Academy of Sciences*, *Management Science*, and *Psychological Science*. Enrique is one of the co-investigators of the *ESRC Network for Integrated Behavioural Sciences* and the National Science Foundation Collaborative Research NSCC/SA project *Behavioural Insights into National Security Issues*.

**Tanya Rosenblat** is an Associate Professor of Economics and Information at the University of Michigan. Dr. Rosenblat is an experimental economist who uses a combination of laboratory and field experiments to study the economics of social (non-market) interactions. She has held appointments at the Institute for Quantitative Social Science at Harvard University, Center for Behavioral Economics at the Federal Reserve Bank of Boston, Institute for Advanced Study (Princeton), National Bureau of Economic Research, California Institute of Technology, Iowa State University of Science and Technology and Wesleyan University. Her PhD is in Economics from Massachusetts Institute of Technology. Her research has been funded by the National Science Foundation and the Russell Sage Foundation and appeared in the *American Economic Review* and the *Quarterly Journal of Economics*. She is a member of the Executive Committee of Economic Science Association and an Associate Editor for Behavioral Economics at *Management Science*.

# PART I
# INTRODUCTION

# ENCOURAGING REPLICATION OF ECONOMICS EXPERIMENTS

Cary A. Deck, Enrique Fatas and Tanya Rosenblat

Experimental economics offers several advantages over more traditional observational approaches in economics. It enables researchers to exogenously manipulate factors of interest. It allows random assignment of observational units. It also allows for replication. Replication *within* a paper is ubiquitous – rarely do experimental papers involve a single observation in a treatment. Replication *across* papers is simultaneously common and uncommon. Many papers include a baseline condition that is *similar* to what a previous paper has done. As argued by Smith (1994, p. 128), "If I do an experiment similar to yours as a baseline control for comparison with a related experiment I intend to perform, I am testing the robustness of your original results using my instructions, subjects and a different experimenter." What is relatively uncommon in experimental economics is direct replication, which many scholars seem to believe is due to a reluctance by journal editors to publish replications.

This volume of *Research in Experimental Economics* was created to create an outlet for replication. When submitted papers were sent to reviewers, the editorial instructions stated that the success or failure of the replication attempt should have no bearing on the recommendation to accept or reject the paper. Rather the decision was based on the faithfulness of the

**Replication in Experimental Economics**
**Research in Experimental Economics, Volume 18, 3–9**

**ISSN: 0193-2306/doi:10.1108/S0193-230620150000018015**

replication attempt, the competence of the analysis, and the degree to which the results had been replicated previously.

Taken together, the papers in this volume reveal the variety of possible approaches to replicating economics experiments as well as some of the issues that researchers must tackle when attempting to replicate existing work. This volume is comprised of four parts. Part I is this introduction. Part II consists of papers comprised solely of direct replication attempts: Eckel, Harwell, and Castillo (2015) and Murphy, Batmunkh, Nilsson, and Ray (2015). Part III includes papers that incorporate replication as a basis for extending the original work in a particular direction: Cox and James (2015), Jiang, Lunawat, and Shapiro (2015), James and Shogren (2015), Zylbersztejn (2015), and Bekkers (2015). Part IV takes a look forward for ways to continue encouraging replication in experimental economics: Maniadis, Tufano, and List (2015).

Eckel et al. (2015) replicate a sequence of four public goods experiments in a sequence of research projects run by PhD students at the University of Texas at Dallas: Fehr and Gachter (2000) on sanctions, Andreoni (1995b) on positive and negative framing, Andreoni (1995a) on confusion, and Isaac and Walker (1988) on MPCR and group size. The comparative statics do confirm the main findings of the original studies, and document an interesting *Texas* effect, as participants exhibit significantly larger contributions to the public good than in the original experiments.

Murphy et al. (2015) attempt to replicate a result from the work by Shang and Croson (2009). In the original paper, phone operators taking donations for a public radio station informed callers of the amount another donor had given. The reference amount was the treatment variable. Shang and Croson (2009) found that providing a higher social reference led to greater donations for new donors. Murphy et al. (2015) follow a similar procedure for the call in campaign of Alaska Public Media. This paper highlights some of the issues that arise when attempting to replicate a field experiment. The actual entities are different although both are US-based public radio stations. The locations are thus different, which means that the characteristics of the participants differ in a myriad of ways and likely to a greater degree than college undergrads at different schools. The replicators are forced to make a decision between using the same dollar amounts, perhaps adjusted to reflect differences in purchasing power, and using the same percentiles (50th, 85th, and 90th) from a different distribution of previous donations. They opt for the latter. The radio stations also had their own idiosyncratic polices, such as Alaska Public Media requiring that callers were asked about becoming "sustainers" who would make monthly payments. Ultimately, Murphy et al. (2015) find no effect from the social reference for new donors. That is, Murphy et al. (2015) do not find the same effect that Shang and Croson (2009) reported for new donors. Murphy et al. (2015) also find that the social reference does not affect repeat donors, which replicates the findings of Shang and Croson (2009) for that group.

Cox and James (2015) conduct new experiments with the centipede game of McKelvey and Palfrey (1992). In this game, players alternate deciding whether to take the lion's share of the current pie or to allow the pie to grow but give decision-making authority. Participants who only worry about their own payment should, using backwards induction, end the game immediately. The original work of McKelvey and Palfrey (1992) found that people did not end the game immediately and only gradually ended the game earlier with repetition. Cox and James (2015) provide two replications of this environment: one that uses the same nominal payoffs and one that uses inflation adjusted payoffs to account for the nearly 25-year differences between the studies. Holding the real stakes fixed, Cox and James (2015) report similar behavior when subjects first play the game and replicate the finding that play does not completely unravel. However, they find that play does unravel faster with repetition than what McKelvey and Palfrey (1992) reported. Cox and James (2015) go on to evaluate how other aspects of the game affect behavior. These extensions include changing the payment of the non-decision maker to zero, limiting decision time and providing a default action, as well as changing the presentation between a tree and a clock format. These factors are similar to ones used by Cox and James (2012) which looked at private payoffs while examining the structural differences between centipede games and Dutch auctions.

Jiang et al. (2015) replicate the investment game of Berg, Dickhaut, and McCabe (1995). The investment game is now widely used to study trust and reciprocity and thus has been replicated directly and indirectly many times. However, the original paper also examined how presenting social history (a description of how another group of subjects played the game) impacted behavior and this aspect of the paper is less commonly considered. Jiang et al. (2015) replicate the social history treatment of Berg et al. (1995) and extend it from two generations to five generations. Using entropy to measure dispersion in behavior, Jiang et al. (2015) find that effect of providing social history is not as robust as in Berg et al. (1995).

James and Shogren (2015) revisit an experiment by List, Berrens, Bohara, and Kerkvliet (2004) showing that voter isolation affects preferences for giving. In the work by List et al. (2004), groups of participants at the University of Central Florida were given money and allowed to vote on whether or not the group should donate money to the Center for Environmental Policy Analysis on campus. People who knew their vote would remain anonymous were less likely to support donation. James and Shogren (2015), running at the University of Wyoming, conduct a similar study using the Wind Energy Research Center on that campus. Beyond the benefactor, James and Shogren (2015) change some design features such as the participation payment and study location (classroom instead of laboratory) and extend the original work by varying the group size. What they find is that the effect of voter isolation holds in larger groups, but not in smaller groups.

Zylbersztejn (2015) studies the behavioral impact of non-verbal feedback in repeated games, separating the signaling and the sanctioning effect of feedback. It does not only replicate previous seminal papers (Masclet, Noussair, Tucker, & Villeval, 2003; Peeters & Vorsatz, 2013), but also documents how non-monetary sanctions improve the provision of the public. It controls both (forward-looking) reputation building and (backward-looking) information effects with a random re-matching mechanism. The intuition is that while both partner matching and stranger matching feedback may work as a non-monetary sanctioning device, only the former allows for strategic signaling. The experimental design neutralizes any confounding effects of the matching protocol on subjects' behavior (see Andreoni & Croson, 2008).

Bekkers (2015) replicates Eckel and Grossman's (2003) analysis of rebate and matching subsidies and supports the greater effectiveness of matching subsidies using a nationally representative sample of the Dutch population. The behavioral rationale for this is that matching schemes do engage more donors because they increase donors' expectations about other donors to participate. Interestingly, Bekkers (2015) shows that matching schemes raise donations among higher educated individuals and large donors.

It is our hope that that this volume will help emphasize the need for replication in experimental economics. Indeed, this is a cause that has been gaining attention recently. The newly created *Journal of the Economic Science Association* explicitly invites replication papers, which we see as very positive. Interestingly, the two papers in this volume of *Research in Experimental Economics* that are purely replication attempts were conducted by established scholars with a group of students as part of a teaching activity. We see this a model that can and should be *replicated.*

As replication becomes more commonplace, it will be important to establish norms and best practices. As alluded to in the above quote from

Smith (1994), "replication" of an economic experiment is never pure. Either the subjects change or their experience does. None of the papers in this volume went so far as to use the same facility as the original study nor were they run at the same times of day or with (deliberately) research assistants of the same sex although any of these factors could influence behavior.[1]

A group of prominent scholars including Colin Camerer, Anna Dreber, Teck Ho, Jürgen Huber, Magnus Johannsson, and Michael Kirchler are currently undertaking replications of 18 papers recently published in *The American Economic Review* or *The Quarterly Journal of Economics*. A novel feature of this endeavor is that their creation of prediction markets where scholars can trade replication shares, assets that pay a positive award if a given study replicates. One implication of creating these markets is that "success" is predefined. According to an April 2, 2015, email distributed to the Economic Science Association listserv:

> A replication [of one of the 18 studies] is defined as a test statistic (using the same analysis in the original paper) with a hypothesis test p-value less than .05. The sample sizes were chosen so that the original result has a 90% chance of replication or higher, given a true effect size of the same magnitude as in the original study.

The final paper in this volume of *Research in Experimental Economics* by Maniadis et al. (2015) discusses a path forward.

## NOTE

1. Smith (1994, p. 128) follows a similar logic as he follows the quote mentioned above with "In effect, I am varying some of the more routine auxiliary hypotheses, and asking if the results are nonetheless indistinguishable."

## REFERENCES

Andreoni, J. (1995a). Cooperation in public-goods experiments: kindness or confusion? *American Economic Review*, *85*, 891–904.

Andreoni, J. (1995b). Warm glow vs. cold prickle: The effect of positive and negative framing on cooperation in experiments. *Quarterly Journal of Economics*, *110*(1), 1–21.

Andreoni, J., & Croson, R. (2008). Partners versus strangers: Random rematching in public goods experiments. In *Handbook of experimental economics results* (Vol. 1, pp. 776–783). Amsterdam: Elsevier.

Bekkers, R. (2015). When and why matches are more effective subsidies than rebates. In C. A. Deck, E. Fatas, & T. Rosenblat (Eds.), *Replication in experimental economics* (Vol. 18).

Research in Experimental Economics. Bingley, UK: Emerald Group Publishing Limited.

Berg, J., Dickhaut, J., & McCabe, K. (1995). Trust, reciprocity, and social history. *Games and Economic Behavior*, *10*, 122–142.

Cox, J. C., & James, D. (2012). Clocks and trees: Isomorphic Dutch auctions and centipede games. *Econometrica*, *80*, 883–903.

Cox, J. C., & James, D. (2015). On replication and perturbation of the McKelvey and Palfrey centipede game experiment. In C. A. Deck, E. Fatas, & T. Rosenblat (Eds.), *Replication in experimental economics* (Vol. 18). Research in Experimental Economics. Bingley, UK: Emerald Group Publishing Limited.

Eckel, C. C., & Grossman, P. J. (2003). Rebate versus matching: Does how we subsidize charitable contributions matter? *Journal of Public Economics*, *87*, 681–701.

Eckel, C., Harwell, H., & Castillo, G. J. G. (2015). Four classic public goods experiments: A replication study. In C. A. Deck, E. Fatas, & T. Rosenblat (Eds.), *Replication in experimental economics* (Vol. 18). Research in Experimental Economics. Bingley, UK: Emerald Group Publishing Limited.

Fehr, E., & Gachter, S. (2000). Cooperation and punishment in public goods experiments. *American Economic Review*, *4*(90), 980–994.

Isaac, R. M., & Walker, M. J. (1988). Group size effects in public goods provision: The voluntary contributions mechanism. *Quarterly Journal of Economics*, *1*(103), 179–200.

James, A. G., & Shogren, J. F. (2015). Revisiting the effect of voter isolation. In C. A. Deck, E. Fatas, & T. Rosenblat (Eds.), *Replication in experimental economics* (Vol. 18). Research in Experimental Economics. Bingley, UK: Emerald Group Publishing Limited.

Jiang, X., Lunawat, R., & Shapiro, B. (2015). The impact of financial histories on individuals and societies: A replication of and extension of Berg et al. (1995). In C. A. Deck, E. Fatas, & T. Rosenblat (Eds.), *Replication in experimental economics* (Vol. 18). Research in Experimental Economics. Bingley, UK: Emerald Group Publishing Limited.

List, J., Berrens, R., Bohara, A., & Kerkvliet, J. (2004). Examining the role of social isolation on stated preferences. *American Economic Review*, *94*(3), 741–752.

Maniadis, Z., Tufano, F., & List, J. A. (2015). How to make experimental economics research more reproducible: Lessons from other disciplines and a new proposal. In C. A. Deck, E. Fatas, & T. Rosenblat (Eds.), *Replication in experimental economics* (Vol. 18). Research in Experimental Economics. Bingley, UK: Emerald Group Publishing Limited.

Masclet, D., Noussair, C., Tucker, S., & Villeval, M. C. (2003). Monetary and non-monetary punishment in the voluntary contributions mechanism. *American Economic Review*, *93*(1), 366–380.

McKelvey, R., & Palfrey, T. R. (1992). An experimental study of the centipede game. *Econometrica*, *60*, 803–836.

Murphy, J. J., Batmunkh, N., Nilsson, B., & Ray, S. (2015). The impact of social information on the voluntary provision of public goods: A replication study. In C. A. Deck, E. Fatas, T. Rosenblat & (Eds.), *Replication in experimental economics* (Vol. 18). Research in Experimental Economics. Bingley, UK: Emerald Group Publishing Limited.

Peeters, R., & Vorsatz, M. (2013). Immaterial rewards and sanctions in a voluntary contribution experiment. *Economic Inquiry*, *51*(2), 1442–1456.

Shang, J., & Croson, R. (2009). A field experiment in charitable contribution: The impact of social information on the voluntary provision of public goods. *The Economic Journal*, *119*, 1422–1439.

Smith, V. (1994). Economics in the laboratory. *Journal of Economic Perspectives*, *8*(1), 113–131.

Zylbersztejn, A. (2015). Nonverbal feedback, strategic signaling, and nonmonetary sanctioning: New experimental evidence from a public goods game. In C. A. Deck, E. Fatas, & T. Rosenblat (Eds.), *Replication in experimental economics* (Vol. 18). Research in Experimental Economics. Bingley, UK: Emerald Group Publishing Limited.

# PART II
# DIRECT REPLICATION

# FOUR CLASSIC PUBLIC GOODS EXPERIMENTS: A REPLICATION STUDY

Catherine C. Eckel, Haley Harwell and José Gabriel Castillo G.

## ABSTRACT

*This paper replicates four highly cited, classic lab experimental studies in the provision of public goods. The studies consider the impact of marginal per capita return and group size; framing (as donating to or taking from the public good); the role of confusion in the public goods game; and the effectiveness of peer punishment. Considerable attention has focused recently on the problem of publication bias, selective reporting, and the importance of research transparency in social sciences. Replication is at the core of any scientific process and replication studies offer an opportunity to reevaluate, confirm or falsify previous findings. This paper illustrates the value of replication in experimental economics. The experiments were conducted as class projects for a PhD course in experimental economics, and follow exact instructions from the original studies and current standard protocols for lab experiments in economics. Most results show the same pattern as the original studies, but in all cases with smaller treatment effects and lower statistical significance,*

**Replication in Experimental Economics**
**Research in Experimental Economics, Volume 18, 13–40**

**ISSN: 0193-2306/doi:10.1108/S0193-230620150000018001**

*sometimes falling below accepted levels of significance. In addition, we document a "Texas effect," with subjects consistently exhibiting higher levels of contributions and lower free-riding than in the original studies. This research offers new evidence on the attenuation effect in replications, well documented in other disciplines and from which experimental economics is not immune. It also opens the discussion over the influence of unobserved heterogeneity in institutional environments and subject pools that can affect lab results.*

**Keywords:** Replication; lab experiment; public goods; experimental economics; research transparency

**JEL classifications:** C92; H41; D64; B40

## INTRODUCTION

Considerable attention has been focused on the problems of publication bias, selective reporting, and the importance of research transparency in the social sciences, especially in recent years. Publication bias occurs because articles with findings that are statistically significant, theoretically interesting, and novel are more likely to be published than null, dull, or replication studies. Selective reporting means that more interesting or novel findings within a study are more likely to be published, and insignificant or non-intuitive results left in a file drawer. The evidence on reported significance levels suggests a serious bias, leading Ioannidis (2005) to assert that "most published research findings are false."[1] His model predicted "rates of wrongness" in medical research of 80% for non-randomized studies, 25% of randomized trials (the "gold standard" for experimental research), and 10% of large-scale randomized trials ("platinum standard"). These correspond roughly to the rates at which medical study results are overturned. In a related observation, Schooler (2011) notes that attempts to repeat studies often result in an apparent decline in treatment effects (such as estimates of drug effectiveness) over time, a phenomenon he has referred to as "cosmic habituation" (Lehrer, 2010), and he calls for a open repository of unpublished results to help combat this problem. Replication seems an important element of the response to demonstrated bias in published results.

Closer to home, a recent controversy over the replicability of research results in social psychology, and in priming research in particular, is highlighted in an open letter from Daniel Kahneman (Yong, 2012), who called on his colleagues in priming research to respond to the criticism.

> I believe that you should collectively do something about this mess. To deal effectively with the doubts you should acknowledge their existence and confront them straight on, because a posture of defiant denial is self-defeating. Specifically, I believe that you should have an association, with a board that might include prominent social psychologists from other field [sic]. The first mission of the board would be to organize an effort to examine the replicability of priming results, following a protocol that avoids the questions that have been raised and guarantees credibility among colleagues outside the field.
>
> —Yong, 2012; supplementary material

The profession was quick to respond. Subsequent efforts led to a special issue of the *Journal of Social Psychology* (2014) on replications, which included, among a number of other papers, a notable effort by the "Many Labs" replication project to replicate 13 major results in the field (Klein et al., 2014). This paper reports the results of experiments conducted in 36 labs with over 6,000 subjects, and finds that 10 effects are robust, one is weak, and two fail to replicate (both priming studies, as it happens).

The flurry of replication in social psychology was inspired by a storm of criticism and skepticism, including the discovery of fraudulent behavior by social psychologists such as Diederik Stapel, Dirk Smeesters, and Lawrence Sanna, but economics research has not come under similar attack. Nevertheless, replicability remains an important aspect of the appeal of experimental research, and is a valuable activity. Although Plott (1982) refers to replication as "the heart of experimental economics," published replication is not commonplace in experimental economics.[2] Published replications are most often found as a first step in a new research project or extension (e.g., Hung & Plott, 2001). In the realm of public goods experiments, one of the most popular "canonical games" in economics (Eckel, 2007), many studies have replicated prior results in somewhat different environments. With this in mind, we set out explicitly to replicate as closely as possible several important, highly cited public goods experiments. Students in a PhD course in Experimental Economics were randomly assigned to four groups, and each group selected a paper to replicate from a list of the most highly cited papers in public goods experimental research (those with at least 200 Google Scholar citations at the time). The papers selected are: Isaac and Walker (1988), Andreoni (1995a, 1995b); and Fehr and Gächter (2000). These papers contain some of the most important results in the study of public goods.

The Voluntary Contribution Mechanism public goods game is a simple environment designed to test theories about social dilemmas. In these games subjects make decisions in groups, typically of size 2–10, and each member of the group must allocate a fixed endowment between an individual investment, which pays \$1 to the decision maker for each \$1 invested, and a public investment, which pays a lower amount (less than \$1 but more than \$1/*n*), termed the Marginal Per Capita Return (or MPCR), to each member of the group. The incentive structure mimics that of a public good, and embodies the classic tension between the social optimum of full contributions to the public good, and individual payoff-maximization, which entails free riding on the contributions of others. From the earliest multi-period studies using this game, results have shown positive levels of contributions that then deteriorate over time in most environments (Ledyard, 1995).

The first of the studies we replicate, Isaac and Walker (1988), tests the effect of changing two factors, the MPCR and group size. In the original study, the authors find that the MPCR has an important impact on contributions, with a move from .3 to .75 increasing average contributions and decreasing the number of free riders. The effect of changing group size from 4 to 10 is weaker.[3] A subsequent experiment tests the effect of groups of size 40 and 100 and finds high levels of public goods production with these very large groups (Isaac, Walker, & Williams, 1994). Holding the MPCR constant, contributions are highest with the largest groups, probably reflecting the total benefit associated with transferring a token to the group account. However with MPCR = .75, the contributions are uniformly high and do not vary significantly with group size. In a comprehensive survey of public goods research, Ledyard (1995) lists MPCR as a factor having a robust positive effect on donations and group size as a factor having a robust zero effect (see his Table 10 and section 3.3 for a discussion of the studies he considers). In a meta-analysis of 27 public goods game experiments, Zelmer (2003) also finds a strong positive effect of MPCR ($p < .0001$), and also finds a positive effect for group size, but with only weak significance ($p < .10$; see Table 3, *p*. 307).

Andreoni (1995a, 1995b) conducted two important variations on the public goods game. In the former, he examined whether the contribution level in the game might be due to confusion, since free riding is a dominant strategy, and found considerable evidence that confusion plays a role. In the second, he explored the effect of positive or negative framing on contribution levels. These two studies constitute the second and third studies selected for replication. Each is described in turn.

In Andreoni (1995a) the experiment consisted of three treatments: *regular*, a standard public good game with groups of size 5 and an MPCR of 0.5; *rank*, where earnings depended only on the subjects' relative payoff in the game and, as a result, there was no payoff to cooperation; and *regrank*, which provided the same information on relative earnings as the *rank* condition, but earnings were determined as in the *regular* condition. He found that approximately half of the total contribution to a public good game could be attributed to kindness and the other half to potential confusion, but that confusion decreased rapidly across the first five rounds, while kindness remained more or less stable. The presence of confused subjects has been confirmed by Houser and Kurzban (2002) and Ferraro and Vossler (2010), but in somewhat different environments. Both of these latter studies involve a comparison between play against real counterparts with play against computerized players, providing a simpler test of confusion than Andreoni (1995a), and both studies also show substantial confusion that declines over time.

Andreoni (1995b) focuses on framing, and its approach is to vary the description of the game in two treatments: a positive-frame description, in which subjects' contributions to the group account generate a positive externality for others in the group; and a negative-frame description in which contributing to the individual account generates a negative externality for each member of the group. The payoff functions are identical in the two treatments; only the description changes. This manipulation leads to substantially lower average contributions and more free riders in the negative-frame treatment. A number of studies have tested the effect of positive and negative framing in variations on the public goods game, but none has attempted to replicate exactly the original study. Sonnemans, Schram, and Offerman (1998) find a similar, but weaker, effect in a binary, step-level public good experiment; however, Bougherara, Denant-Boemont, and Masclet (2011) show that framing has an even more powerful effect in a provision-point mechanism public good setting. Willinger and Ziegelmeyer (1999) replicate the effect of framing differences in a public good game with an interior Nash equilibrium; they find that while positive framing leads to above-equilibrium contributions, negative framing leads subjects to play the Nash equilibrium. Park (2000) explores heterogeneity in responses to positive and negative framing and show stronger effects for subjects who have and individualistic value orientation, in contrast to this with a cooperative value orientation. In a somewhat different environment, Shanley and Grossman (2007) show that subjects are more willing to contribute to a public good than to refrain from contributing to a "public bad" that, in essence, deteriorates a public good; they find no difference in the number of free riders. Grossman and Eckel (2015) find no effect of positive and negative framing in a "real donation" dictator game study, where donations go to a charitable organization producing public goods in the field.

Finally, the fourth study, Fehr and Gächter (2000), is one of the first to test the effect of sanctions on public goods contributions. This study adds an additional stage to each round of the public good game in which subjects can, at a cost to themselves, reduce the payoffs of other members of their group. The effect of punishment is tested in a "partners" setting, where groups are stable across rounds of the game, and a "strangers" setting, where groups are reconstituted each round. Punishment is very effective in enhancing contributions. Since their paper was published, there have been many studies (too many to survey here) testing various aspects of the structure and effectiveness of punishment institutions. (Chaudhuri, 2011) provides a useful survey of research up to that point. In a concluding statement he notes that the opportunity to punish in such games generally increases contributions, but given the cost of punishment, does not always enhance efficiency. Besides the issue with efficiency, he suggests two additional cautions. First, he notes that punishment itself is a second-order public good, and could require punishment of the non-punishers (free riders on punishment) as well as the free riders on contributions. Second, most studies show some anti-social punishment, which appears to be counter-punishment of high contributors by low contributors. Several recent studies highlight the importance of preventing anti-social punishment for the effectiveness of punishment institutions (DeAngelo & Gee, 2015; Faillo, Grieco, & Zarri, 2013; Hauser, Nowak, & Rand, 2014).

Our replication studies yield two main results. First, although our data replicate the patterns in the original data, the treatment effects are consistently smaller than in the original studies. In some cases, the results of the original study cannot be replicated, in that the treatment effect sizes fall below statistical significance. Second, all studies show a positive "Texas effect," with the subjects at the University of Texas at Dallas consistently contributing higher amounts to the public good than in the original studies. While Texans are more generous than the subjects in the original studies, it is important to note that the level of contributions is within the range found in cross-cultural comparisons.

## EXPERIMENTAL DESIGN AND PROCEDURE

All of the experiments were conducted in a way that mimicked as closely as possible the original conditions of the experiments, with one exception: To ensure comparability across replications, all experiments were computerized. When possible, the original program was obtained from the authors, but otherwise the games were programmed using z-tree (Fischbacher, 2007). Payment was adjusted to reflect current norms of payment for subjects: all received a $5 show-up fee, and average earnings were $13–$19.

Subjects for all experiments were recruited using ORSEE (Greiner, 2004), and sessions took place at the Center for Behavioral and Experimental Economic Science (CBEES) at the University of Texas at Dallas, in April and May 2012.[4] A summary of the titles and treatments is included in Table 1.

### *Isaac and Walker, Group Size Effects in Public Goods Provision*

The 2x2 factorial design includes two treatments, varying the group size ($n$) from 4 to 10; and the MPCR from .3 to .75, as in the original study. The design also varies the total endowment of tokens for each individual across groups, with an eye to keeping the maximum earnings relatively constant across treatments. The endowment varied in the four person treatment from 62 tokens with the lower .3 MPCR, to 25 in the .75 MPCR rounds. The endowment also varied with the MPCR in the 10-group treatment. Subjects were endowed with 25 tokens in the low .3 MPCR treatment to 10 tokens in the high .75 MPCR rounds. One treatment is played for 10 rounds in stable groups, followed by a surprise restart and a second 10 rounds with a different MPCR. For the 4-person groups, sessions included 12 subjects (three groups of 4) in anonymous groups. For the 10-person groups, each session was a single group.[5] Subjects were randomly assigned to a computer station; dividers ensured that decisions were anonymous. The replication used z-tree to code the experiment, where the original authors used PLATO as their computerized platforms.

***Table 1.*** Original and Replication Design Comparison.

| | Isaac and Walker | | Andreoni Kindness/ Confusion | | Andreoni Warm Glow/ Cold Prickle | | Fehr and Gächter | |
|---|---|---|---|---|---|---|---|---|
| | 1988 | Repl. | 1995 | Repl. | 1995 | Repl. | 2000 | Repl. |
| Show-up fee | n/a | $5 | n/a | $5 | n/a | $5 | 15CHF ($9) | $5 |
| Subjects | 84 | 84 | 120 | 120 | 80 | 60 | 112 | 48 |
| Sessions | 12 | 12 | 2 | 12 | 2 | 12 | 5 | 5 |
| Average earnings ($) | n/a | 15.00 | 8.68 | 13.92 | 8.24 | 15.28 | 41CHF ($25) | 18.70 |
| Time (minutes) | n/a | n/a | 50 | 30 | 50 | 41 | 120 | 70 |
| Program | Plato | z-Tree | None | z-Tree | None | z-Tree | z-Tree | z-Tree |

Subjects completed self-paced instructions that were the same as in the original study. At the end of each round subjects were shown a summary of group contributions and their own group and individual earnings from the round. These were totaled after the final round and converted to dollars at the rate of $0.005 per token. The average individual earnings, including a $5 show up fee, were about $15.

After subjects completed the first treatment of 10 rounds, they were informed they would participate in another 10 rounds. To account for sequencing effects, the treatment order was counterbalanced, with six sessions having high then low MPCR, and six the reverse.[6]

### *Andreoni, Kindness or Confusion*

This experiment replicates the three treatments of the original design: named *regular*, *rank*, and *regrank*, which vary the earnings calculation (based on the standard game or based on rank) and information (standard or standard + rank), as explained above. The design is between-subjects, with each session consisting of a single treatment played in groups of 5 persons for 10 rounds, with 2 anonymous groups per session. The endowment is 60 tokens, and the MPCR is 0.5. The experiment was coded in z-tree; the original study was conducted by hand.

The *regular* condition is the standard game. In the *rank* condition, payoffs are calculated based only on the rank of the subject's earnings in the game. This change preserves the dominant strategy of the basic game: the strategy to achieve the highest earnings rank is to be the free rider of the group. However, it also removes any incentives to contribute since others don't benefit from one's contribution. Effectively, this treatment "makes a zero-sum payoff game out of a standard public goods game" (Andreoni, 1995a). Any remaining contributions could be attributed to confusion. Because these two treatments differ in both the earnings structure and the information given to the subjects about rank, the third treatment, *regrank*, removes the confound in the design by using the standard payoff calculation but providing rank information. Earnings were about $14 including a $5 show-up fee.

### *Andreoni, Warm Glow versus Cold Prickle*

The design consists of two treatments: a standard game, which is a "positive" frame mentioning the positive externality of a contribution to the group account on the other group members, and a negatively framed game, which emphasizes the negative externality associated with a contribution to the individual account. The design is between-subjects, with each session consisting of a single treatment played in groups of five persons for 10 rounds, with two anonymous, rematched groups per session.[7] The endowment is 60 tokens, and the MPCR is 0.5. Like the Andreoni (1995a) study, the experiment is coded in z-tree; the original study was conducted by hand. Earnings were about $15 including the $5 show-up fee.

The positive frame had standard instructions. In the negative treatment subjects were again given 60 tokens in each round and were told they could allocate them however they wish between the two accounts. However, subjects were also given automatic earnings totaling 120 additional tokens each round. They were told that investing in the individual account offered a one-for-one payoff for the subject but also led to a deduction of one-half a token from the earnings of all the other group members.[8]

### *Fehr and Gächter, Cooperation and Punishment*

The 2x2 design varies two elements of the game: whether there is punishment, and whether groups are stable (partners) or reconstituted each period (strangers). The group composition is between-subjects, and the punishment element is within-subjects. In each session subjects in groups of four first complete 10 rounds of punishment, followed by 10 rounds without, or vice versa. Each session consists of 12 subjects (three groups of four). Subjects received an endowment of 20 tokens and the MPCR of the public account is 0.4. The experiment is based on a translation of the original study and is programmed in z-tree. Earnings were about $19, including the show up fee.

In the no-punishment condition, the standard game is played for 10 rounds. The punishment treatment adds a second stage where subjects can simultaneously punish one another, at a cost to themselves. The cost of punishing is the same as in the original paper:

| | | | | | | | | | | | |
|---|---|---|---|---|---|---|---|---|---|---|---|
| Punishment points | 0 | 1 | 2 | 3 | 4 | 5 | 6 | 7 | 8 | 9 | 10 |
| Cost of punishment | 0 | 1 | 2 | 4 | 6 | 9 | 12 | 16 | 20 | 25 | 30 |

## RESULTS

### *Isaac and Walker (1988)*

Tests of treatment differences using aggregate results from Isaac and Walker (1988) and the Texas replication are shown in Table 2. The levels of contributions and proportion of "strong free riders" (subjects with contributions less than 1/3 of their endowments) are shown. (Additional material is available in the data appendix, available on request, including results by period.) For example, the first row of Table 2 tests whether low and high MPCR generate different outcomes in groups of four players; the second performs the same test for groups of 10. While in the original study, high MPCR leads to significantly higher contributions for both group sizes, we are unable to replicate that result. The impact of group size is never significant in our data, though it is in Isaac and Walker for low MPCR. However, we do replicate three out of four comparisons on the number of strong free riders in each treatment.

Figs. 1(a) and (b) show the pattern of results over time in the Isaac/Walker and replication data. In both, the left panel graphs the Isaac/Walker data and the right the Texas replication. It is easy to see that the treatment differences are much weaker in the Texas data – neither changes in MPCR or group size appear to have substantial impacts on contribution levels.

***Table 2.*** Treatment Effects: Contributions and "Strong" Free Riders. Non-Parametric Tests (Isaac & Walker, 1988 and Replication).

| H0: Mean Equality (In Both Studies, $N=12$ for 4L and 4H; $N=40$ for 10H and 10L) | | Isaac and Walker | | TX-Replication | |
|---|---|---|---|---|---|
| | | *z* | *p*-Value | *z* | *p*-Value |
| *Contributions* | *4L vs. 4H* | −4.107 | 0.000 | −0.691 | 0.490 |
| | *10L vs. 10H* | −3.200 | 0.001 | −1.218 | 0.223 |
| | *4H vs. 10H* | 0.451 | 0.652 | 0.010 | 0.992 |
| | *4L vs. 10L* | −4.666 | 0.000 | 0.000 | 1.000 |
| *Number of strong free riders* | *4L vs. 4H* | 5.101 | 0.000 | 2.959 | 0.003 |
| | *10L vs. 10H* | 2.888 | 0.004 | 1.115 | 0.265 |
| | *4H vs. 10L* | −5.428 | 0.000 | −5.158 | 0.000 |
| | *4H vs. 10H* | −5.379 | 0.000 | −5.158 | 0.000 |
| | *4L vs. 10L* | −4.933 | 0.000 | −4.272 | 0.000 |

*Note*: Wilcoxon rank-sum test performed at individual level. *p*-Value for $|z|$. Tests include the complete 20 periods (two sequences) of both data sets. Strong free riders are those with contributions less than 1/3 of endowment.

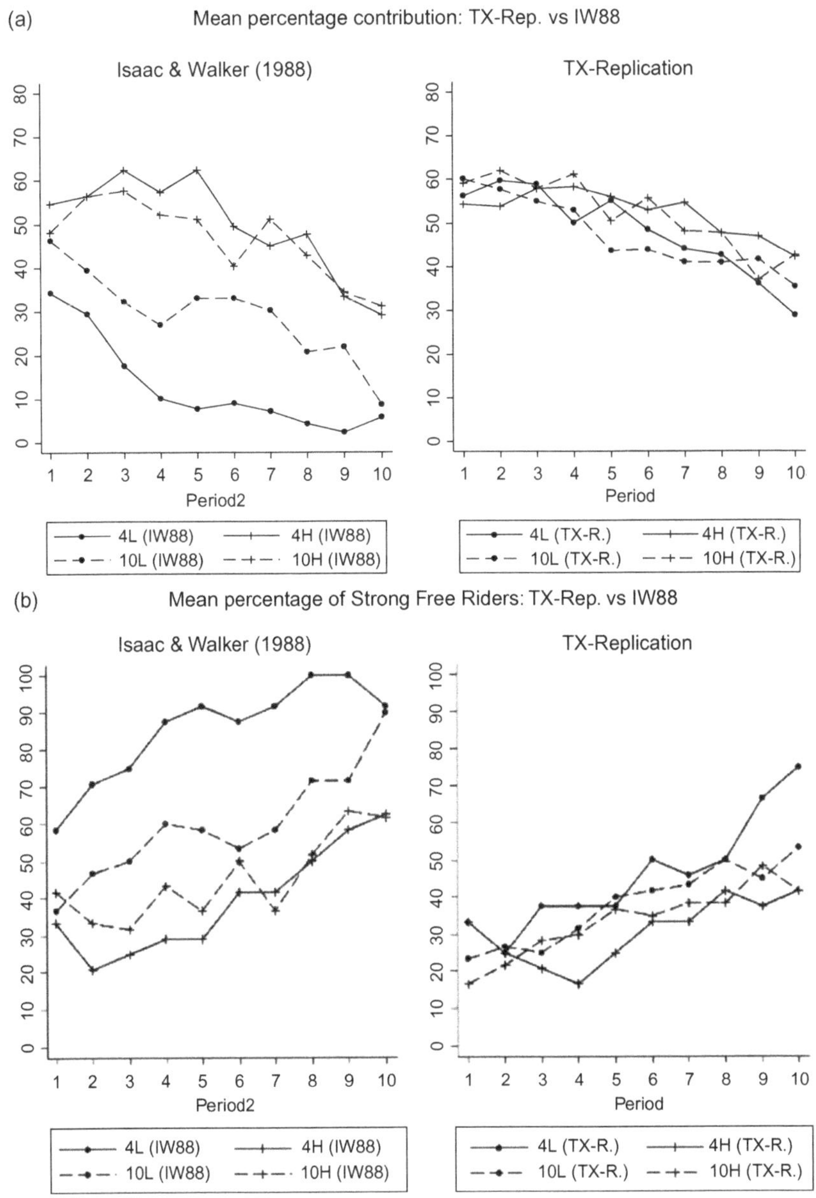


*Fig. 1.* (a) Comparisons of Mean Contributions: Isaac and Walker (1988) with Texas Replications. (b) Comparisons of Strong Free Riders: Isaac and Walker (1988) with Texas Replications.

Regression analysis on the percent contributions shows significant treatment effects for both MPCR and group size in the pooled data (Table 3). The first model also reveals a negative interaction term indicating the treatment effect is different for the larger groups. Notably, there is a positive, significant coefficient on a dummy variable indicating the Texas replication data. In the second model, we add Texas interaction effects for the treatment variables. The treatment effects are clearly different in the replication; indeed, the interaction effects indicate that there are no meaningful treatment effects for the Texas data – the interaction coefficients offset the main effects. Furthermore, we find no significant group size effect and incidence of the MPCR on free riders is significant only conditional on a small

***Table 3.*** Determinants of Contributions (Isaac & Walker 1988 and Replication).

| | Dependent Variable: Percentage Contributions | | |
|---|---|---|---|
| | Model 1 | Model 2 | Model 3 |
| Texas (replication = 1) | 0.139*** | 0.353** | 0.353** |
| | (0.0385) | (0.149) | (0.150) |
| MPCR (high = 1) | 0.207*** | 0.370*** | 0.276*** |
| | (0.0435) | (0.0161) | (0.0131) |
| Group size (high = 1) | 0.0785 | 0.165*** | 0.114*** |
| | (0.0755) | (0.0198) | (0.0176) |
| MPCR*group size | −0.0961*** | −0.197*** | −0.132*** |
| | (0.0263) | (0.0188) | (0.0189) |
| Texas*MPCR | | −0.325*** | −0.224*** |
| | | (0.0448) | (0.0505) |
| Texas*Group Size | | −0.172 | −0.433*** |
| | | (0.143) | (0.0960) |
| Texas*MPCR*group size | | 0.202*** | 0.102 |
| | | (0.0519) | (0.0577) |
| Other's average contribution: OAC($t-1$) | | | 0.276*** |
| | | | (0.0304) |
| Texas*OAC($t-1$) | | | −0.201*** |
| | | | (0.0336) |
| Constant | 0.235*** | 0.128*** | 0.0810*** |
| | (0.0634) | (0.00826) | (0.00433) |
| $N$ (observations) | 3360 | 3360 | 3192 |
| $N$ (subjects) | 168 | 168 | 168 |
| $r^2$ | 0.0761 | 0.0970 | 0.128 |

*Notes*: Random effects estimation. Robust standard errors clustered at group level in parentheses. Dependent variable = Percentage contributions: contribution/endowment by experiment. Significance at: **5% level, ***1% level.

group size. The results are unchanged when we control for the contributions of others, using a variable capturing the others' average contribution in the group for the previous round (OAC($t-1$)).

The data for the first replication study fail to fully replicate the original study's results. While the pattern of results is consistent with the original study, the effect sizes are quite different.

### *Andreoni (1995a): Kindness or Confusion*

Tests for aggregate treatment effects for Andreoni (1995a) are summarized in Table 4 below. This table shows that we were unable to find a significant difference between treatments *reg* versus *regrank* in the replication data. The difference between *regrank* and *rank* is marginally significant, and we do see significant differences between the *reg* and *rank* treatments. Per-period averages for each treatment, which follow a similar pattern to the original results with contributions for *regular* > *regrank* > *rank*, and free riders for *regular* < *regrank* < *rank*, but smaller differences between the treatments (details available on request).

The pairwise comparisons for the treatments for all 10 periods are shown in Fig. 2(a)–(c). (We illustrate these separately because putting all in one figure is overly cluttered.) The *regular* condition shows very similar behavior in the two studies. The *regrank* and *rank* conditions show higher contributions in Texas than the original.

In the original paper, confusion is measured as the percent of subjects contributing positive amounts in the *rank* condition, and kindness as the relative difference between *regrank* and *rank*. In both treatments, subjects

***Table 4.*** Treatment Effects: Contributions and Pure Free Riders. Non-Parametric Tests (Andreoni, 1995a and Replication).

| H0: Mean Equality ($n=40$ for Both Studies) | | Andreoni (1995a) | | TX-Replication | |
|---|---|---|---|---|---|
| | | $z$ | $p$-Value | $z$ | $p$-Value |
| *Contributions* | Regular vs. Regrank | 3.766 | 0.000 | 1.473 | 0.141 |
| | Regrank vs. Rank | −3.602 | 0.008 | −1.638 | 0.101 |
| | Regular vs. Rank | 5.091 | 0.000 | 2.995 | 0.003 |
| *Number of pure free riders* | Regular vs. Regrank | −1.817 | 0.069 | −1.705 | 0.088 |
| | Regrank vs. Rank | 2.880 | 0.004 | 0.492 | 0.622 |
| | Regular vs. Rank | −3.602 | 0.000 | −1.971 | 0.049 |

*Note*: Wilcoxon rank-sum test performed at individual level. $p$-value for $|z|$.

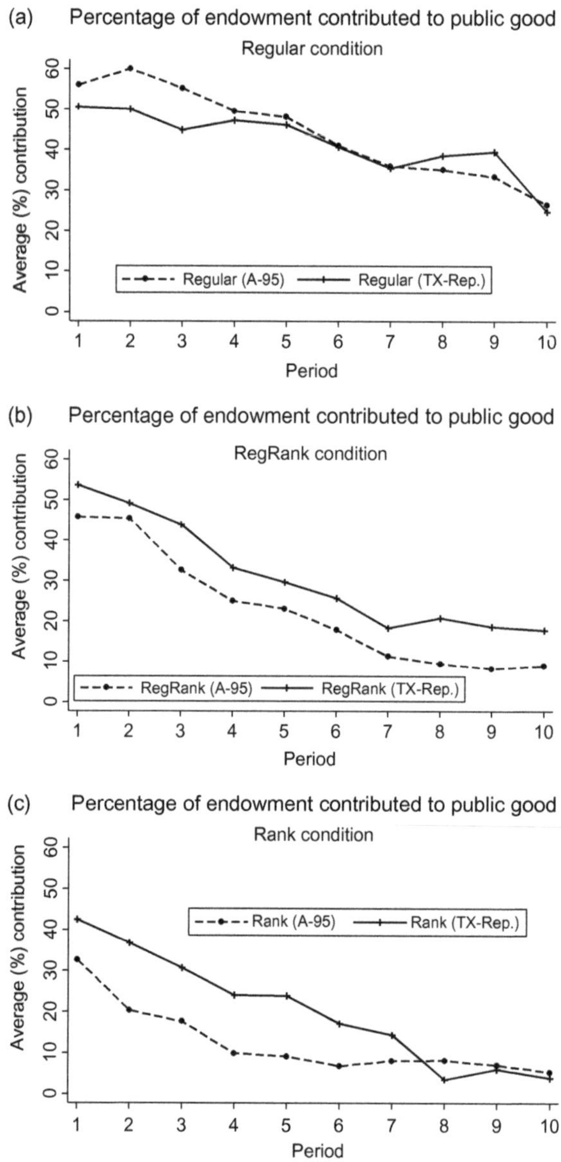


*Fig. 2.* Comparisons of Andreoni (1995a) with Texas Replications.

receive information about rank, but only in the latter treatment does rank alone determine earnings. Andreoni argues that the comparison between these two provides a clean measure of confusion. In the Texas data, confusion appears to play a larger role, as contributions are somewhat higher in the *rank* treatment as compared with the original.[9] While the level of confusion declines over time in both data sets, the levels of confusion are higher in the Texas data (71% of the subjects) and the levels of kindness lower (10% of the subjects). However, the proportion of subjects that could not be classified as motivated by either criterion is larger in Andreoni's data.

Table 5 shows a regression analysis, pooling the original and replication data. The first and second models are the same except in the way the standard errors are calculated (robust versus clustered on the group level). Using the more conservative second model, this illustrates the treatment

***Table 5.*** Determinants of Contributions (Andreoni, 1995a and Replication).

| | Dependent Variable: Contributions (Tokens per Period: 60) | | | |
|---|---|---|---|---|
| | Model 1 | Model 2 | Model 3 | Model 4 |
| Texas (replication = 1) | −1.375 | −1.375 | −0.450 | −3.120 |
| | (1.639) | (3.231) | (1.597) | (2.336) |
| Treatment 2: Regrank | −18.93*** | −18.93*** | −9.503** | −10.59* |
| | (1.418) | (2.123) | (2.575) | (1.508) |
| Treatment 3: Rank | −12.78*** | −12.78** | −6.899* | −7.62** |
| | (1.435) | (3.933) | (2.643) | (1.94) |
| Texas*Regrank | 6.025*** | 6.025 | 2.289 | 3.941* |
| | (2.032) | (4.011) | (3.062) | (1.457) |
| Texas*Rank | 6.352*** | 6.352 | 2.948 | 3.963 |
| | (2.152) | (5.374) | (2.926) | (1.886) |
| Other's avg. contribution − OAC($t-1$) | | | 0.381*** | 0.339*** |
| | | | (0.0570) | (0.00912) |
| Texas*OAC($t-1$) | | | | 0.0741 |
| | | | | (0.101) |
| Constant | 26.46*** | 26.46*** | 11.63** | 13.17*** |
| | (1.152) | (2.059) | (2.970) | (1.111) |
| *N* (observations) | 2400 | 2400 | 2160 | 2160 |
| *N* (subjects) | 240 | 240 | 240 | 240 |
| $r^2$ | 0.104 | 0.104 | 0.206 | 0.207 |

*Notes*: Random effects estimation. Robust standard errors in parentheses clustered at group level for models in columns 2, 3, and 4. Excluded treatment: Regular.
Significance at: *10% level, **5% level, ***1% level.

effects in the original study, with the *regrank* treatment reducing contributions from the regular treatment, and rank reducing it still further. However, the Texas interactions partially offset the treatment effects, showing that in the replication data the effect sizes are smaller. In some cases these fall below statistical significance. Texas contributions are higher in both the *rank* and *regrank* treatments.

In sum, we see the same pattern of results in the replication as in the original data, but the effect sizes are smaller. Confusion appears to be somewhat higher in the replication. Texas contributions are higher in two out of the three treatments.

### *Andreoni (1995b): Warm Glow versus Cold Prickle*

The results for Andreoni (1995b) are summarized in Table 6. The treatment difference between the positive and negative frame are successfully replicated, with significant differences in both contributions and the number of free riders between the two treatments. (The tables from the original study are replicated in the supplementary material, available on request, which also includes contributions for each period of play.)

Fig. 3 illustrates the findings and shows the pattern of behavior over time. The original data (dashed lines) clearly show the treatment effect and the Texas data replicate the effect. Notably, the replication shows a substantially higher level of contributions for both treatments.[10]

Regression analysis in Table 7 confirms these results. Using robust or clustered standard errors, the treatment effect for positive frame (in contrast to negative frame) is positive and statistically significant. There is a pronounced Texas effect in the data, with Texans contributing on average about twice as much as in the original study (number of free riders is half of

***Table 6.*** Treatment Effects: Contributions and Free Riders. Non-Parametric Tests. (Andreoni, 1995b and Replication).

| | | Andreoni (1995b) | | TX-Replication | |
|---|---|---|---|---|---|
| | | $z$ | $p$-Value | $z$ | $p$-Value |
| Contributions | Positive vs. negative frame | −3.508 | 0.0005 | −2.611 | 0.0090 |
| Free riders | Positive vs. negative frame | 3.791 | 0.0002 | 3.121 | 0.0018 |

*Note*: Wilcoxon rank-sum test performed at individual level. $p$-value for $|z|$. Statistics for the original study are calculated from the public data set and slightly differ from those reported in the paper.

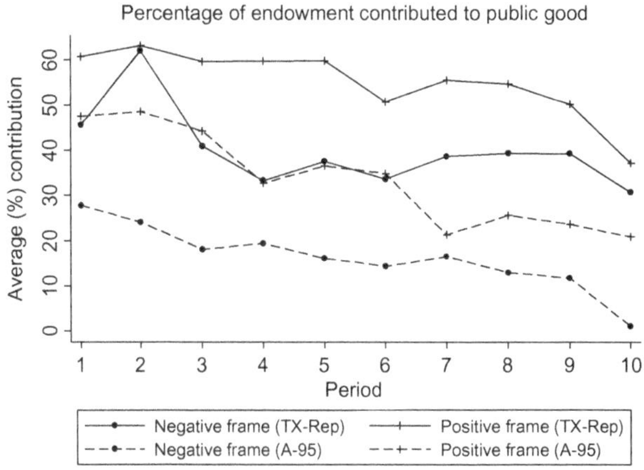


*Fig. 3.* Comparisons of Contributions: Andreoni (1995b) with Texas Replications.

***Table 7.*** Determinants of Contributions (Andreoni, 1995b and Replication).

| | Dependent Variable: Contributions (Tokens per Period: 60) | | | |
|---|---|---|---|---|
| | Model 1 | Model 2 | Model 3 | Model 4 |
| Texas (replication = 1) | 14.31*** | 14.31*** | 10.58*** | 9.607** |
| | (1.578) | (0.661) | (0.527) | (2.675) |
| Treatment (positive frame) | 10.43*** | 10.43** | 7.22*** | 7.504** |
| | (1.350) | (1.825) | (1.041) | (1.516) |
| Texas*treatment | −1.408 | −1.408 | −1.063 | −1.610 |
| | (2.218) | (1.982) | (1.407) | (2.381) |
| Other's avg. contribution: OAC($t-1$) | | | 0.223** | 0.202* |
| | | | (0.0473) | (0.0802) |
| Texas*OAC($t-1$) | | | | 0.0412 |
| | | | | (0.117) |
| Constant | 9.720*** | 9.720*** | 5.757** | 6.053** |
| | (0.853) | (0.887) | (1.399) | (1.545) |
| *N* (observations) | 1400 | 1400 | 1260 | 1260 |
| *N* (subjects) | 140 | 140 | 140 | 140 |
| $r^2$ | 0.146 | 0.146 | 0.175 | 0.175 |

*Notes*: Random effects estimation. Robust standard errors in parentheses clustered at group level for models 2, 3, and 4.
Significance at: *10% level, **5% level, ***1% level.
Excluded treatment: Negative frame.

the original study). Controlling for the contributions of others shows a familiar "matching" effect, with higher contributions by others associated with higher own contributions (Croson, 2007). The interactions effects in the fourth model show that there is no significant difference in the treatment effect between Texas and the original study, and that the Texans do not respond differently to the information about prior contributions of others.

In sum, the treatment effect of a positive versus a negative frame on identical public goods experiments is successfully replicated, with one key difference: subjects in Texas make substantially higher contributions than in the original study.

### *Fehr and Gächter (2000)*

The results for the Fehr and Gächter (2000) paper largely replicate the original study. Recall that the treatments vary group matching (Strangers vs. Partners), punishment (No-Punishment vs. Punishment), and order. Table 8 shows the tests on the average treatment effects for the replication, alongside the same tests with the original data. In all cases the treatment effect is replicated in the Texas data, though treatment effect sizes are smaller ($p < .0002$).

Fig. 4 shows the aggregate effects for the strangers treatment. Panel A shows the results when 10 rounds with punishment precede 10 without and vice versa for panel B. When punishment is first, the Texas data are very close to the original, when punishment is removed, the decay in contributions is slower for the Texas data. When no-punishment is first, the level of contributions in the Texas data is considerably higher, but the treatment

***Table 8.*** Treatment Effects: Average Contributions. Non-Parametric Tests (Fehr & Gächter, 2000 and Replication).

| | No-Punishment vs. Punishment | | | |
|---|---|---|---|---|
| | F&G 2000 | | TX-Replication | |
| | $z$ | $p$-Value | $z$ | $p$-Value |
| *Strangers* | 7.161 | 0.0000 | 3.959 | 0.0001 |
| *Partners* | 2.803 | 0.0051 | 1.992 | 0.0464 |

Two-sample Wilcoxon matched pairs test (signed-rank test) performed for differences at individual level for the strangers treatments (i.e., each observation is an average contribution per subject) and at the group level for the partners treatments. $p$-value for $|z|$. Results for F&G 2000 calculated from original data set.

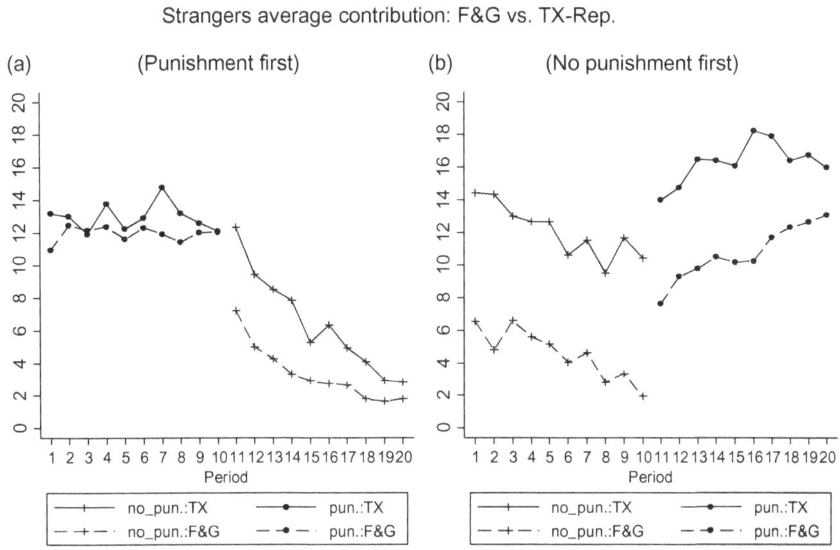


*Fig. 4.* Comparisons of Strangers Contributions: Fehr and Gächter (2000) with Texas Replications.

effect is replicated. (The supplementary appendix, available on request, contains detailed data on the average contributions and last-period contributions.)

We next turn to the partners treatment. Fig. 5 shows the aggregate data when punishment rounds are played first (panel A) and when no-punishment rounds are played first. In panel A, as in the strangers treatments, the Texas data for punishment rounds matches the original data, and the decline is slower once punishment is removed. In panel B, the no-punishment rounds again are above the original; the punishment rounds are slightly below.

We then conduct regression analysis with the pooled data from the original and the replication. These are shown in Table 9. Once again, there is a pronounced Texas effect, with Texans contributing on average approximately three tokens (about 40%) more than in the original study. The regression confirms that punishment increases contributions in both pairings – partners and strangers – but is larger in the partners treatments. In addition, the interaction between Texas and the punishment treatment is negative and significant, indicating a smaller (though still significant) treatment effect in the replication.

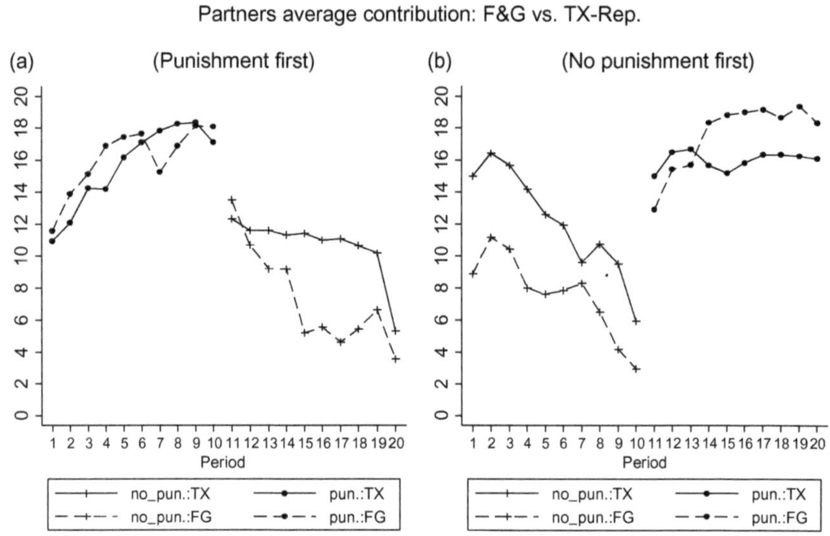


*Fig. 5.* Comparisons of Partners Contributions: Fehr and Gächter (2000) with Texas Replications.

***Table 9.*** Determinants of Contributions (Fehr & Gächter, 2000 and Replication).

| | Dependent Variable: Contributions (20 Tokens per Period) | | |
|---|---|---|---|
| | Model 1 | Model 2 | Model 3 |
| Texas (replication = 1) | 4.788*** | 4.788** | 2.334** |
| | (0.402) | (1.371) | (0.725) |
| Treatment (punishment) | 8.095*** | 8.095*** | 6.148*** |
| | (0.295) | (0.435) | (0.270) |
| Texas*treatment | −3.628*** | −3.628** | −2.392** |
| | (0.514) | (1.349) | (0.902) |
| Partner (=1) | 3.245*** | 3.245** | 1.547 |
| | (0.359) | (1.133) | (0.772) |
| Treatment*partner | 0.843* | 0.843 | 0.657 |
| | (0.462) | (1.025) | (0.469) |
| Other's avg. contribution: OAC($t-1$) | | | 0.497*** |
| | | | (0.0249) |
| Received punishment points: RPP ($t-1$) | | | −0.756*** |
| | | | (0.185) |
| Constant | 3.919*** | 3.919*** | 1.548*** |
| | (0.209) | (0.390) | (0.326) |
| *N* (observations) | 3200 | 3200 | 3040 |
| *N* (subjects) | 160 | 160 | 160 |
| $r^2$ | 0.338 | 0.338 | 0.454 |

*Notes*: Random effects estimation. Robust standard errors in parenthesis clustered at group level for models in columns 2 and 3.
Significance at: *10% level, **5% level, ***1% level.

From this analysis we conclude that the treatments are replicated. In all conditions, punishment enhances contributions. However, there are two clear differences from the original study. The first is that there is a "Texas effect" – with subjects in Texas contributing significantly more on average than in the original study. Second, the treatment effects are smaller than in the original study, they remain statistically significant only on the strangers' treatment.

## CONCLUSION

We conducted replications of four highly cited research papers in public goods experiments. For the most part we are able to replicate results from prior studies. In the Isaac and Walker (1988) study, we do not find significant treatment effects on average contributions of changes in the MPCR in either group size (4 or 10), but we do find significant effects (echoing the original study) on the number of strong free riders. For Andreoni (1995a) we replicate the pattern of results showing a strong presence of confusion, and, if anything, show a higher proportion of confusion-based decisions in the replication than in the original. For Andreoni (1995b) we find a strong effect of framing on decision making, as in the original study. Finally, for Fehr and Gächter (2000), we replicate the result that punishment increases contributions in both partners and strangers pairings (statistical significance evident only on the strangers' pairing), though our effect sizes are smaller than in the original study.

Our results show two main findings. First, the most notable feature of our results is that in all cases the treatment effects are smaller in the replication than in the original study. Table 10 summarizes this finding. The left side of the table shows, for each treatment combination, the effect size for the original study (difference between treatments), compared with the effect size for the replication. In all cases these amounts are directly comparable between the original and the replication, as we replicated the parameters in the original studies. The third column calculates the difference in treatments. In every case but one (Andreoni, *regrank* – *rank*), the difference carries a negative sign, indicating that the replication effects are smaller than in the original studies. The largest differences are for the Isaac and Walker paper and the smallest for Andreoni's warm glow/cold prickle framing study. This decline in treatment effects was predicted by Schooler (2011), who noted a pronounced decline in treatment effects across a wide variety

***Table 10.*** Treatment Effects and the Texas Effect. Differences in the Average Contributions (as Percent of Endowment) to the Public Good.

| | Treatment Effects | | | The Texas Effect | |
|---|---|---|---|---|---|
| | Original | Replication | Diff. (rep. – original) | Diff. (rep. –original) | |
| *Isaac-Walker (1988)* | | | | | |
| 4H–4L | 36.99 | 4.46 | −32.53 | 4L | 35.26 |
| 10H–10L | 17.28 | 4.96 | −12.32 | 4H | 2.73 |
| 10L–4L | 16.47 | −0.78 | −17.25 | 10L | 18.01 |
| 10H–4H | −3.24 | −0.28 | 2.96 | 10H | 5.69 |
| *Andreoni K/C (1995a)* | | | | | |
| Reg-Regrank | 21.30 | 10.71 | −10.59 | Reg | −2.29 |
| Regrank-Rank | 10.24 | 10.79 | 0.55 | Regrank | 8.30 |
| Reg-Rank | 31.54 | 21.50 | −10.04 | Rank | 7.75 |
| *Andreoni W/G (1995b)* | | | | | |
| WG-CP | 17.38 | 15.03 | −2.35 | Positive | 21.51 |
| | | | | Negative | 23.86 |
| *Fehr-Gächter (2000)* | | | | | |
| P-NP (strangers) | 39.0 | 26.9 | −12.15 | NP (strangers) | 27.80 |
| P-NP (partners) | 47.5 | 22.0 | −25.5 | P (strangers) | 15.65 |
| | | | | NP (partners) | 19.50 |
| | | | | P (partners) | −6.00 |

*Notes*: Units reported are the average percentage contribution relative to the endowment in each study for consistency.

of types of studies. It seems as if economics experiments are not immune from this effect. The pattern is consistent with publication bias, where more significant results are more likely to be selected for publication. However, we are quick to say we do not in any way wish to cast aspersions on the original studies, which were conducted with state-of-the-art methods and show interesting and important results.

Our second main result concerns the higher level of contributions in the subject pool used for the replication studies. The right side of Table 10 makes this point (The Texas Effect). Here, for every treatment we have calculated the difference between the *level* of contributions in the original study and the *level* of contributions in the replication. The last column of the table contains this value, with a positive sign indicating higher contributions in the replication. In all but two cases, contributions are higher in the replication. That is, across all studies we observe a Texas effect, with Texans giving at higher levels than in the original study.

The magnitude of the Texas effect is within the range of cross-cultural variation in public goods game experiments reported in several previous studies. For example, Henrich et al. (2001) report considerable variation in the results of experiments conducted in 15 small scale societies around the world (see Henrich et al. (2004) for details). While their main analysis focuses on the range of play in the ultimatum game, in six of the communities data were also collected using the public goods game. As in the ultimatum game, considerable variation is seen, ranging from zero contributions among the hunter-gatherer Machiguenga, to about 60% contributions among the Orma. More relevant for our work is by Herrmann, Thöni, and Gächter (2008), who find substantial cross-cultural variation in mean contribution rates in public good games with and without punishment. These games were conducted with undergraduate students in 16 countries. Without punishment the range of contributions was from an average of 4.9 out of 20 tokens (24.5%) in Melbourne, Australia, and 11.5 tokens (57.5%) in Copenhagen. With punishment contributions range from 5.7 tokens (28.5%) in Athens to 18 (90%) in Boston. Gächter and Herrmann (2009) discuss how variations in reciprocity and contribution can be attributed to specific aspects of culture.[11] Contributions in our sample from the University of Texas at Dallas are at the high end of these ranges.

Of course within the United States there is considerably less variation in the factors that their study explores. Nevertheless, their strong evidence of a cultural difference in cooperative behavior is consistent with our findings of what appears to be a "Texas Effect." A cultural norm seems to be producing higher levels of contribution to the public good. We can only speculate as to the causes. The demographics of the subject pool at UT Dallas may differ systematically from those of the earlier studies, but this is impossible to test as the original studies contain no demographic information. The other studies were conducted for the most part at large state universities (Arizona, Wisconsin). UT Dallas is not the flagship university in Texas (that would be UT Austin). In comparison, UT Dallas is smaller, it has the highest entering SAT scores among Texas universities (because of its focus on computer science and engineering), and has a larger fraction of ethnically Indian and Chinese students (48% – many of these are second-generation), reflecting the predominance of tech-related industries in the north Dallas area where the university is located. This is in comparison to Arizona and Wisconsin, where about 10% of students were minorities in 1995. We are unaware of any studies showing systematic differences in contributions associated with these factors.

Higher giving in the game might simply be due to higher levels of personal generosity among Texans. It is also possible that confusion plays a role, as evidenced by the somewhat higher level of confusion shown in the replication of Andreoni (1995a). However, it is not unreasonable to think that Texans might be more generous, stereotypes to the contrary. (Whenever we tell colleagues that we have a pronounced Texas effect in our public goods games, we ask them to guess which way the effect goes, and inevitably they assume the worst of Texans.)

Texas is strongly conservative, with Republicans controlling all statewide offices, dominating the state legislature, and holding both US Senate seats and 25/36 seats in the US House of Representatives. The Republican candidate has won the presidential race in all elections in the last three decades. The major Texas cities, including Austin, Dallas, Houston, and San Antonio, usually support Democrats, while their suburbs are heavily Republican. UT Dallas draws largely from Dallas and its suburbs, and so has roughly equal proportions of students who identify themselves as Republican or Democratic. We did not collect political information in this study.

Claims are often made of greater giving by conservatives. For example, Brooks (2007) claims that conservatives make 30% higher donations of money than liberals (even controlling for income), give more blood and even donate more of their time. Texas ranks 13th among the states in donations among those who itemize tax deductions, according to the Chronicle of Philanthropy (August 19, 2012), with donations of 5.1% of discretionary income. However, most studies using national surveys show little difference in giving by political orientation (Anderson, Mellor, & Milyo, 2005; Margolis & Sances, 2013), and most lab experiments show no difference in secular giving by party affiliation (Fowler & Kam, 2007) or religiosity (Eckel & Grossman, 2004). In sum, we are unable to explain the Texas effect in our data, and leave it to other studies to replicate the phenomenon.

## NOTES

1. Statistician Sterling (1959) was the first to note a pattern consistent with selective reporting, when he observed that 97% of all published studies in psychology that had statistically significant results confirmed the effect they sought to test.

2. A new project was recently announced that will replicate 18 experimental papers in economics published between 2011 and 2014 in the *American Economic Review* and the *Quarterly Journal of Economics*. This project also included prediction markets, which are asset markets designed to predict which studies will be replicated successfully. The markets are now closed, but results have not been announced. (See http://sciencepredictionmarkets.com).

3. This paper replicates an earlier study, Isaac, Walker, and Thomas (1984). Isaac and Walker (1988) conduct little in the way of statistical analysis. Ashley,

Ball, and Eckel (2010) reanalyze their data using regression analysis and are able to show that group size does have a significant effect on contributions, but only when the MPCR is low.

4. This Center was closed in August 2012.

5. The CBEES has a 12-station lab, which limited the number of subjects in each session.

6. We use all 20 rounds of data in our analysis. In the original paper, Isaac and Walker state that they are only using the final 10 periods of data to be conservative in their tests, but we believe they may have used all 20 rounds as well based on our reanalysis of their data.

7. The original study ran four groups per session, with rematching. Limits of our lab meant that we could only run two groups per session.

8. Holcomb and Eckel (2014), one of the class papers, includes these data, as well as two additional treatments varying the language in the instructions.

9. Tables 8 and 9 in the supplementary Appendix, available on request, show the measures derived by Andreoni in his Table 2. Appendix Table 8 shows the difference in the percent of subjects contributing zero by treatment, and shows that the pattern of behavior is similar in the replication data.

10. Table 12 in the supplementary appendix, available on request, contains tests of statistical significance for the Texas effect.

11. Substantial cross cultural differences are also documented in Gächter and Herrmann (2009), Buchan et al. (2009), Kachelmeier and Shehata (1997), among other studies.

## ACKNOWLEDGMENTS

Thanks to the staff of the Center for Behavioral and Experimental Economic Science (CBEES), especially Wendy Mak Lee, who guided the students through the programming of the experiments; Eric McLester, who handled the IRB applications and communication; Sheheryar Banuri, who was TA for the course; and the many undergraduate research assistants who helped conduct the sessions: Ravi Hanumara, Tanushree Jhunjhunwala, Nick Lafferty, Nathan Eacret, Anthony Tantillo, Addison Ziegler. Zhengzheng Wang and Billur Aksoy assisted with manuscript preparation. Funding was provided by CBEES and the Negotiations Center at UT Dallas.

The course papers on which this paper is based are as follows:

Manca, Simone, Sabrina Ren, and Noha Sobi, "A Replication of: Group Size Effects in Public Good Provision: The Voluntary Contribution Mechanism."

George, Justin, and Parneet Pahwa, "A Replication of Cooperation in Public-Goods Experiment: Kindness or Confusion."

Holcomb, Alex, Victoria Obiadeze, and Shidi Wang, "Framing, It Really Does Matter."

Alzahrani, Ahmed, Haley Harwell, Caitlin McKillop, and Zhengzheng Wang "A Replication of 'Cooperation and Punishment in Public Goods Experiments."

## REFERENCES

Anderson, L. R., Mellor, J. M., & Milyo, J. (2005). Do liberals play nice? The effects of party and political ideology in public goods and trust games. In J. Morgan (Ed.), *Advances in Applied Microeconomics*. Bradford, UK: Emerald Group Publishing.

Andreoni, J. (1995a). Cooperation in public-goods experiments: Kindness or confusion? *American Economic Review*, *85*, 891–904.

Andreoni, J. (1995b). Warm glow vs. cold prickle: The effect of positive and negative framing on cooperation in experiments. *Quarterly Journal of Economics*, *110*(1), 1–21.

Ashley, R., Ball, S., & Eckel, C. (2010). Motives for giving: A reanalysis of two classic public goods experiments. *Southern Economic Journal*, *77*(1), 15–26.

Bougherara, D., Denant-Boemont, L., & Masclet, D. (2011). Cooperation and framing effects in provision point mechanisms: Experimental evidence. *Ecological Economics*, *70*(6), 1200–1210.

Brooks, A. C. (2007). *Who really cares: The surprising truth about compassionate conservatism – America's charity divide – who gives, who do*. Basic Books. New York: Basic Books, c2006.

Buchan, N. R., Grimalda, G., Wilson, R., Brewer, M., Fatas, E., & Foddy, M. (2009). Globalization and human cooperation. *Proceedings of the National Academy of Sciences*, *106*(11), 4138–4142.

Chaudhuri, A. (2011). Sustaining cooperation in laboratory public goods experiments: A selective survey of the literature. *Experimental Economics*, *14*(1), 47–83.

Croson, R. T. A. (2007). Theories of commitment, altruism, and reciprocity: Evidence from linear public goods games [article]. *Economic Inquiry*, *45*(2), 199–216. doi:10.1111/j.1465-7295.2006.00006.x

DeAngelo, G., & Gee, L. K. (2015). Should we centralize detections and sanctions? *P*eer vs. centralized detection and punishment in the provision of public goods. Working Paper. Tufts University. unpublished manuscript.

Eckel, C. C. (2007). Economic games for social scientists. In M. Webster & J. Sell (Eds.), *Laboratory experiments in the social sciences* (pp. 497–515). Oxford: Elsevier.

Eckel, C. C., & Grossman, P. J. (2004). Giving to secular causes by the religious and nonreligious: An experimental test of the responsiveness of giving to subsidies. *Nonprofit and Voluntary Sector Quarterly*, *33*(2), 271–289. doi:10.1177/0899764004263423

Faillo, M., Grieco, D., & Zarri, L. (2013). Legitimate punishment, feedback, and the enforcement of cooperation. *Games and Economic Behavior*, *77*(1), 271–283.

Fehr, E., & Gächter, S. (2000). Cooperation and punishment in public goods experiments. *American Economic Review*, *90*(4), 980–994.

Ferraro, P. J., & Vossler, C. A. (2010). The source and significance of confusion in public goods experiments. *The B.E. Journal of Economic Analysis & Policy*, *10*(1). Article 53.

Fischbacher, U. (2007). Z-tree: Zürich toolbox for readymade economic experiments. *Experimental Economics*, *10*(2), 171–178.

Fowler, J. H., & Kam, C. D. (2007). Beyond the self: Social identity, altruism, and political participation. *Journal of Politics*, *69*(3), 813–827.

Gächter, S., & Herrmann, B. (2009). Reciprocity, culture and human cooperation: Previous insights and a new cross-cultural experiment. *Philosophical Transactions of the Royal Society B: Biological Sciences*, *364*(1518), 791–806.

Greiner, B. (2004). An online recruitment system for economic experiments. In K. Kremer & V. Macho (Eds.), *Forschung und wissenschaftliches Rechnen (Research in scientific computing)* (Vol. GWDG Bericht 63, pp. 79–93). Göttingen: Ges. für Wiss. Datenverarbeitung.

Grossman, P. J., & Eckel, C. C. (2015). Giving versus taking for a cause. *Economics Letters*, *132*, 28–30.

Hauser, O. P., Nowak, M. A., & Rand, D. G. (2014). Punishment does not promote cooperation under exploration dynamics when anti-social punishment is possible. *Journal of Theoretical Biology*, *360*, 163–171.

Henrich, J., Boyd, R., Bowles, S., Camerer, C., Fehr, E., & Gintis, H. (2004). *Foundations of human sociality: Economic experiments and ethnographic evidence from fifteen small-scale societies*. Oxford: Oxford University Press.

Henrich, J., Boyd, R., Bowles, S., Camerer, C., Fehr, E., Gintis, H., & McElreath, R. (2001). In search of homo economicus: Behavioral experiments in 15 small-scale societies. *American Economic Review*, *91*(2), 73–78.

Herrmann, B., Thöni, C., & Gächter, S. (2008). Antisocial punishment across societies. *Science*, *319*(5868), 1362–1367.

Houser, D., & Kurzban, R. (2002). Revisiting kindness and confusion in public goods experiments. *American Economic Review*, *92*(4), 1062–1069.

Hung, A. A., & Plott, C. R. (2001). Information cascades: Replication and an extension to majority rule and conformity-rewarding institutions. *American Economic Review*, *91*(5), 1508–1520.

Ioannidis, J. P. (2005). Why most published research findings are false. *PLoS Medicine*, *2*(8), e124.

Isaac, R. M., Walker, J., & Thomas, S. (1984). Divergent expectations on free riding: An experimental examination of possible explanations. *Public Choice*, *43*, 113–149.

Isaac, R. M., & Walker, J. M. (1988). Group size effects in public goods provision: The voluntary contributions mechanism. *Quarterly Journal of Economics*, *103*(1), 179–200.

Isaac, R. M., Walker, J. M., & Williams, A. W. (1994). Group size and the voluntary provision of public goods: Experimental evidence utilizing large groups. *Journal of Public Economics*, *54*, 1–36.

Kachelmeier, S. J., & Shehata, M. (1997). Internal auditing and voluntary cooperation in firms: A cross-cultural experiment. *Accounting Review*, *72*(3), 407–431.

Klein, R. A., Ratliff, K. A., Vianello, M., Adams, R. B., Jr., Bahník, Š., Bernstein, M. J., & Brumbaugh, C. C. (2014). Investigating variation in replicability. *Social Psychology*, *45*(3), 142–152.

Ledyard, J. O. (1995). Public goods: A survey of experimental research. In J. H. Kagel & A. E. Roth (Eds.), *The handbook of experimental economics*. Princeton, NJ: Princeton University Press.

Lehrer, J. (2010). The truth wears off. The New Yorker, 13, 52.

Margolis, M., & Sances, M. (2013). Who really gives? Partisanship and charitable giving in the United States. Partisanship and Charitable Giving in the United States. August 9, 2013.

Park, E.-S. (2000). Warm-glow versus cold-prickle: A further experimental study of framing effects on free-riding. *Journal of Economic Behavior & Organization*, *43*(4), 405–421.

Plott, C. R. (1982). Industrial organization theory and experimental economics. *Journal of Economic Literature*, *20*(4), 1485–1527.

Schooler, J. (2011). Unpublished results hide the decline effect. *Nature*, *470*(7335), 437.

Shanley, J., & Grossman, P. J. (2007). Paradise to parking lots: Creation versus maintenance of a public good. *The Journal of Socio-Economics*, *36*(4), 523–536.

Sonnemans, J., Schram, A., & Offerman, T. (1998). Public good provision and public bad prevention: The effect of framing. *Journal of Economic Behavior & Organization*, *34*(1), 143–161. Retrieved from http://dx.doi.org/10.1016/S0167-2681(97)00042-5

Sterling, T. D. (1959). Publication decisions and their possible effects on inferences drawn from tests of significance – or vice versa. *Journal of the American Statistical Association*, *54*(285), 30–34.

Willinger, M., & Ziegelmeyer, A. (1999). Framing and cooperation in public good games: An experiment with an interior solution. *Economics Letters*, *65*(3), 323–328.

Yong, E. (2012). Nobel laureate challenges psychologists to clean up their act. *Nature News*, 490.

Zelmer, J. (2003). Linear public goods experiments: A meta-analysis. *Experimental Economics*, *6*(3), 299–310.

# THE IMPACT OF SOCIAL INFORMATION ON THE VOLUNTARY PROVISION OF PUBLIC GOODS: A REPLICATION STUDY

James J. Murphy, Nomin Batmunkh, Benjamin Nilsson and Samantha Ray

## ABSTRACT

*Shang and Croson (2009) found that providing information about the donation decisions of others can have a positive impact on individual donations to public radio. In this study, we attempted to replicate their results, but found* no *evidence that social information affected donation decisions. However, most of our donors were renewing members, a group which Shang and Croson also found was not influenced by social information.*

**Keywords:** Charitable giving; field experiment; philanthropy; public goods; social information; social comparison

**JEL classifications:** C93; D64; H41

**Replication in Experimental Economics**
**Research in Experimental Economics, Volume 18, 41–50**

**ISSN: 0193-2306/doi:10.1108/S0193-230620150000018002**

# INTRODUCTION

The desire to conform to social norms can be a strong influence on human behavior (Bernheim, 1994; Festinger, 1954). Failure to adhere to these norms can lead to moral costs (Levitt & List, 2007), and possibly monetary (Ostrom, Walker, & Gardner, 1992; Yamagishi, 1986) or non-monetary sanctions (Masclet, Noussair, Tucker, & Villeval, 2003). Conformity can also arise as a means of acquiring information by observing and imitating the behavior of others. Models of conformity can explain cooperation in laboratory and field environments, including social dilemmas (e.g., Bardsley & Sausgruber, 2005; Carpenter, 2004; Velez, Stranlund, & Murphy, 2009), employee recognition (Bradler, Dur, Neckermann, & Non, 2014), information cascades (Goeree & Yariv, 2015), and pressing a "Like" button on Facebook (Egebark & Ekström, 2011). Social comparisons may lead to conformity if the behavior of others serves as a reference point for individual decisions (Zafar, 2011), and it has been used effectively to influence behavior in a number of domains, including energy conservation (Allcott, 2011), water conservation (Ferraro, Miranda, & Price, 2011; Ferraro & Price, 2013), hotel towel use (Goldstein, Cialdini, & Griskevicius, 2008; Schultz, Khazian, & Zaleski, 2008), and preferences for environmental goods (Carlsson, García, & Löfgren, 2010).

With the rapid growth in charitable giving (List, 2011), there has been an increase in attention to mechanisms that are effective at fundraising, including public revelation of decisions and the use of social comparisons. In laboratory public goods experiments, public revelation of individual decisions tends to increase cooperation (Andreoni & Petrie, 2004; List, Berrens, Bohara, & Kerkvliet, 2004; Lopez, Murphy, Spraggon, & Stranlund, 2012; Rege & Telle, 2004; Spraggon, Sobarzo, & Stranlund, 2015), and observability of donations by third parties has also yielded higher donations in field experiments (Alpízar & Martinsson, 2013; Soetevent, 2005). Providing social information about the behavior of others has improved cooperation in laboratory public goods games (Zafar, 2011), dictator games (Cason & Mui, 1998), and ultimatum games (Bohnet & Zeckhauser, 2004). In a mail fundraising campaign at a Swiss university, Frey and Meier (2004) found a small positive effect of social information on the decision to donate. Similarly, in a field experiment soliciting donations for a Costa Rican national park, Alpízar and Martinsson (2010) found that informing visitors about common donation amounts led to conformity with the social reference.

The focus of this paper is an attempt to replicate the results of Shang and Croson (2009) who found that some social comparisons can be effective in increasing donations to public radio. During the June and September 2013 on-air fundraising campaign for an anonymous public radio station, Rachel Croson and Jen Shang incorporated a series of social comparison treatments that are described in three papers (Croson & Shang, 2008, 2013; Shang & Croson, 2009). Since the studies were all part of the same pair of pledge drives, the procedures are identical. The key differences among the papers are the subsets of data used and the focus of the analysis. Individuals who called to make a donation were randomly assigned to either a social information condition or a control group. When volunteers answered phones to accept the donation, they read the following script: "Hello, STATION_NAME member line. Are you a new member or a renewing member of STATION_NAME? We had another member, they contributed $X. How much would you like to pledge today?" where $X was the social information provided to the caller. To determine the values of $X, they reviewed the donations from the previous year's campaign. Shang and Croson (2009) reported the results from using the 50th ($75), 85th ($180), and 90th ($300) percentiles; Croson and Shang (2013) used the 95th ($600) and 99th ($1000) percentiles. The combined results of these two papers suggested that social comparisons for public radio donations were most effective at increasing donations when drawn from the 90th to 95th percentiles. At the 99th percentile, presumably the comparison was perceived as unrealistic and was therefore too high to be effective.

Croson and Shang (2008) used the phone campaign data for only the renewing members and combined it with additional data from a mail campaign. They tested whether the difference between the social information condition ($X) and the individual's most recent donation affected the change in the current donation. They found that the change in donations moved in the direction of the social information. Those individuals for whom the social information was greater than their previous donation averaged a $12 increase in giving, whereas those with a lower social information averaged a $24 reduction in contributions.

## EXPERIMENTAL DESIGN

To the extent possible, we followed the procedures described in Shang and Croson (2009). Our field experiment was conducted during the Fall pledge drive (October 15–24, 2014) for Alaska Public Media, based in Anchorage. During the pledge drive, radio listeners were encouraged to call and make a donation. As in their study, we had three social information treatments using data from the 50th, 85th, and 90th percentiles of the previous two Fall pledge drives ($120, $240, and $360, respectively); there was also a no information control. That these amounts occurred in $120 increments was convenient because the station was making an effort to encourage listeners to become "sustaining members" with regular monthly payments.

Phones were answered by volunteers who used the following script: "Thank you for calling Alaska Public Media. Would you like to become a sustainer?" Depending upon the response and treatment, the volunteer read (or omitted in the control): "Another member donated [$120, $240, or $360]. How much would you like to donate?"[1] The pledge drive did not include any thank you gifts or incentives that were based on the amount donated. Unlike Shang and Croson, our volunteers were not blind to the treatment when answering the phone. The scripts were read from proprietary software that had limited customization options. As each call was answered, the software immediately presented the volunteer with a new script that included a randomly assigned treatment.

## RESULTS

The Fall 2014 Alaska Public Media pledge drive received substantially fewer phone donations than the two previous Fall pledge drives, which averaged about 1,500 individual phone donations per campaign. This was the result of two primary factors: (1) a large shift away from phone to online donations and (2) a substantial increase in the number of "sustaining members" who automatically donate by credit card every month and no longer need to call during the pledge drive. Of the 628 donations received, 287 were online and therefore could not be included in the analysis. We also excluded 24 donations of $1000 or more that were made by phone (Table 1). These were all renewing members who had made similar large donations in the past. For the 317 observations that were included in the analysis, donations averaged about $150 with little difference among the treatments. Pairwise comparisons of treatments using nonparametric Wilcoxon rank-sum tests confirmed this conclusion: all tests failed to reject the hypothesis that the treatments and the control were drawn from populations with the same distribution.

***Table 1.*** Distribution of Member Types.

| | Pledge < $1000 | | | Pledge ≥ $1000 | | |
|---|---|---|---|---|---|---|
| Treatment | Renewal | New | Total | Renewal | New | Total |
| Control | 119 | 17 | 136 | 10 | 0 | 10 |
| $120 | 46 | 14 | 60 | 6 | 0 | 6 |
| $240 | 56 | 7 | 63 | 4 | 0 | 4 |
| $360 | 47 | 11 | 58 | 4 | 0 | 4 |
| Online | 220 | 51 | 271 | 15 | 1 | 16 |
| Total | 488 | 100 | 588 | 39 | 1 | 40 |

***Table 2.*** Pledge Amounts.

| | (1) OLS | (2) Robust |
|---|---|---|
| Constant | 97.647*** | 67.521*** |
| | (30.98) | (19.52) |
| $120 (50th percentile) | 2.832 | 5.660 |
| | (19.03) | (11.99) |
| $240 (85th percentile) | −8.578 | −8.360 |
| | (19.21) | (12.10) |
| $360 (90th percentile) | 0.680 | 7.754 |
| | (19.49) | (12.28) |
| Renewing | 35.352* | 37.492*** |
| | (18.99) | (11.96) |
| Sustainer (installment) | 90.835*** | 98.418*** |
| | (18.77) | (11.82) |
| *N* | 317 | 317 |

Models include fixed effects for date.
Statistical significance: ***$p < 0.01$; *$p < 0.10$.

Table 2 presents two regression models that used the same approach as Shang and Croson (their Table 2). The first is a standard OLS regression, and the second is a robust regression which controls for outliers. The models included dummy variables for the three social information treatments. We also controlled for sustaining members (same as the installment variable in Shang and Croson) and for renewing members. The models also included fixed effects for the date of the transaction. For confidentiality reasons, we did not have access to any personal information other than an individual's donation history and cannot control for gender.

For both of our models, none of the coefficients for the social information treatments were statistically significant. A test of the hypothesis that all three social information coefficients were jointly equal to zero cannot be rejected ($p = 0.654$). Consistent with Shang and Croson, the 50th percentile was not expected to be significant; since this is an upward comparison for half the callers and downward for the other half, the two effects should cancel out. However, the absence of a social information effect for the 90th percentile differs from Shang and Croson.

To reconcile the contrasting results, we note that a large majority of our phone pledges were from renewing members (85%). Only 49 of the 317 donations in our analysis were from new members. This was substantially different from Shang and Croson who reported about 40% of donations from renewing members. They suggested that social information is more likely to have an effect in ambiguous circumstances, which would be more germane to new donors. With a roughly even split of new and renewing members, they were able to test this conjecture by rerunning their regression models using only new donors and again using only renewing members (their Table 3). They found that, as conjectured, the social information effect was limited to new donors. Social information at the median value had no effect on new donations, but it was significant at the 85th and 90th percentiles. Their renewing members, on the other hand, were not influenced by the comparisons; none of the social information conditions were statistically significant for this group.

Following Shang and Croson's approach, we reran our regressions separately for new and renewing members (Table 3). Consistent with Shang and Croson, we do not observe a treatment effect for renewing donors.

***Table 3.*** Pledge Amounts by New and by Renewing Members.

| | (1)<br>OLS-New | (2)<br>Robust-New | (3)<br>OLS-Renewing | (4)<br>Robust-Renewing |
|---|---|---|---|---|
| Constant | 242.602*** | 100.000* | 123.676*** | 102.905*** |
| | (84.73) | (52.65) | (26.89) | (17.33) |
| $120 (50th percentile) | 11.558 | −5.415 | −3.723 | 4.154 |
| | (44.28) | (27.51) | (21.42) | (13.81) |
| $240 (85th percentile) | 23.737 | −37.370 | −15.689 | −8.293 |
| | (53.28) | (33.11) | (20.78) | (13.39) |
| $360 (90th percentile) | −14.439 | −44.186 | −4.917 | 12.326 |
| | (47.48) | (29.50) | (21.95) | (14.15) |
| Sustainer (installment) | 71.059* | 84.649*** | 97.655*** | 111.237*** |
| | (37.22) | (23.12) | (22.01) | (14.19) |
| *N* | 49 | 49 | 268 | 268 |

Models include fixed effects for date.
Statistical significance: ***$p < 0.01$; *$p < 0.10$.

However, we also do not observe a treatment effect for new donors, which is contrary to Shang and Croson. This could be due to the relatively small number of new donors ($n = 49$) in our data. Therefore, our results are consistent with one of Shang and Croson's results. Renewing members' donation decisions do not appear to be influenced by social information. However, the low number of new donors in our data prevents us from reliably testing whether this group is affected.

## CONCLUSIONS

In many domains, providing information about the behavior of others has been effective at affecting individual decisions. In a series of papers, Shang and Croson draw on this research and provide social information in an attempt to increase donations for a public radio station. They found that social information can be effective, but only over a limited range. Specifically, their results suggested that social information was most effective for new donors when the social comparison is drawn from the 90th to 95th percentile of previous donations. Our study attempted to replicate their results. We found no evidence that donations were responsive to comparison amounts. This could be because most of our donors were renewing members, who Shang and Croson also found were not responsive to social comparisons. It is also possible that differences in the subject pools (primarily Anchorage residents vs. supporters of an anonymous East Coast public radio station) may explain the contrasting results.

## NOTE

1. Our script closely followed that of Shang and Croson, except that our station wanted to also include the offer to become a sustainer.

## ACKNOWLEDGMENTS

We are particularly grateful to Ann Alquist of Alaska Public Media for her enthusiastic support of this study. We also thank Steve Lindbeck, Torrie Allen, Melissa Walker, and the many volunteers who answered phones for making this study possible. Jen Shang and Rachel Croson were extremely helpful and supportive of this attempt to replicate their original study. Kyle Borash modified the software to randomly assign scripts. Samantha Luban provided outstanding research assistance.

# REFERENCES

Allcott, H. (2011). Social norms and energy conservation. *Journal of Public Economics*, *95*(9–10), 1082–1095. doi:10.1016/j.jpubeco.2011.03.003

Alpízar, F., & Martinsson, P. (2010). *Don't tell me what to do, tell me who to follow*! (No. EfD DP 10-16). Retrieved from http://www.rff.org/rff/documents/EfD-DP-10-16.pdf

Alpízar, F., & Martinsson, P. (2013). Does it matter if you are observed by others? Evidence from donations in the field. *Scandinavian Journal of Economics*, *115*(1), 74–83. doi:10.1111/j.1467-9442.2012.01744.x

Andreoni, J., & Petrie, R. (2004). Public goods experiments without confidentiality: A glimpse into fund-raising. *Journal of Public Economics*, *88*(7–8), 1605–1623. doi:10.1016/S0047-2727(03)00040-9

Bardsley, N., & Sausgruber, R. (2005). Conformity and reciprocity in public good provision. *Journal of Economic Psychology*, *26*(5), 664–681. doi:10.1016/j.joep.2005.02.001

Bernheim, B. D. (1994). A theory of conformity. *Journal of Political Economy*, *102*(5), 841–877. doi:10.1086/261957

Bohnet, I., & Zeckhauser, R. (2004). Social comparisons in ultimatum bargaining. *Scandinavian Journal of Economics*, *106*(3), 495–510. doi:10.1111/j.1467-9442.2004.00372.x

Bradler, C., Dur, R., Neckermann, S., & Non, A. (2014). *Employee recognition and performance: Employee recognition and performance: A field experiment* (No. IZA DP No. 8311). Germany: Bonn. Retrieved from http://papers.ssrn.com/sol3/papers.cfm?abstract_id=2256482

Carlsson, F., García, J. H., & Löfgren, Å. (2010). Conformity and the demand for environmental goods. *Environmental and Resource Economics*, *47*(3), 407–421. doi:10.1007/s10640-010-9385-2

Carpenter, J. P. (2004). When in Rome: Conformity and the provision of public goods. *The Journal of Socio-Economics*, *33*(4), 395–408. doi:10.1016/j.socec.2004.04.009

Cason, T., & Mui, V. (1998). Social influence in the sequential dictator game. *Journal of Mathematical Psychology*, *42*, 248–265. doi:10.1006/jmps.1998.1213

Croson, R., & Shang, J. (Yue). (2008). The impact of downward social information on contribution decisions. *Experimental Economics*, *11*, 221–233. doi:10.1007/s10683-007-9191-z

Croson, R., & Shang, J. (2013). Limits of the effect of social information on the voluntary provision of public goods: Evidence from field experiments. *Economic Inquiry*, *51*(1), 473–477. doi:10.1111/j.1465-7295.2012.00468.x

Egebark, J., & Ekström, M. (2011). *Like what you like or like what others like? Conformity and peer effects on Facebook*. IFN Working Paper No. 886. Stockholm, Sweden. Retrieved from http://papers.ssrn.com/sol3/papers.cfm?abstract_id=1948802

Ferraro, P. J., Miranda, J. J., & Price, M. K. (2011). The persistence of treatment effects with norm-based policy instruments: Evidence from a randomized environmental policy experiment. *American Economic Review*, *101*(3), 318–322.

Ferraro, P. J., & Price, M. K. (2013). Using nonpecuniarystrategies to influence behavior: Evidence from a large-scale field experiment. *The Review of Economics and Statistics, 95*(1), 64–73.

Festinger, L. (1954). A theory of social comparison processes. *Human Relations, 7*, 117–140. doi:10.1177/001872675400700202

Frey, B. S., & Meier, S. (2004). Social comparisons and pro-social behavior: Testing "Conditional Cooperation" in a field experiment. *American Economic Review, 94*(5), 1717–1722.

Goeree, J. K., & Yariv, L. (2015). Conformity in the lab. *Journal of the Economic Science Association, 1*(1), 15–28. doi:10.1007/s40881-015-0001-7

Goldstein, N. J., Cialdini, R. B., & Griskevicius, V. (2008). A room with a viewpoint: Using social norms to motivate environmental conservation in hotels. *Journal of Consumer Research, 35*(3), 472–482. doi:10.1086/586910

Levitt, S. D., & List, J. A. (2007). What do laboratory experiments measuring social preferences reveal about the real world. *Journal of Economic Perspectives, 21*(2), 153–174. doi:10.1257/jep.21.2.153

List, J. A. (2011). The market for charitable giving. *Journal of Economic Perspectives, 25*(2), 157–180. doi:10.1257/jep.25.2.157

List, J. A., Berrens, R. P., Bohara, A. K., & Kerkvliet, J. (2004). Examining the role of social isolation on stated preferences. *American Economic Review, 94*(3), 741–752. doi:10.1257/0002828041464614

Lopez, M. C., Murphy, J. J., Spraggon, J. M., & Stranlund, J. K. (2012). Comparing the effectiveness of regulation and pro-social emotions to enhance cooperation: Experimental evidence from fishing communities in Colombia. *Economic Inquiry, 50*(1), 131–142. doi:10.1111/j.1465-7295.2010.00344.x

Masclet, D., Noussair, C., Tucker, S., & Villeval, M.-C. (2003). Monetary and nonmonetary punishment in the voluntary contributions mechanism. *American Economic Review, 93*(1), 366–380. doi:10.1257/000282803321455359

Ostrom, E., Walker, J., & Gardner, R. (1992). Covenants with and without a sword: Self-governance is possible. *The American Political Science Review, 86*(2), 404. doi:10.2307/1964229

Rege, M., & Telle, K. (2004). The impact of social approval and framing on cooperation in public good situations. *Journal of Public Economics, 88*(7–8), 1625–1644. doi:10.1016/S0047-2727(03)00021-5

Schultz, W. P., Khazian, A. M., & Zaleski, A. C. (2008). Using normative social influence to promote conservation among hotel guests. *Social Influence, 3*(1), 4–23. doi:10.1080/15534510701755614

Shang, J., & Croson, R. (2009). A field experiment in charitable contribution: The impact of social information on the voluntary provision of public goods. *The Economic Journal, 119*, 1422–1439. doi:10.1111/j.1468-0297.2009.02267.x

Soetevent, A. R. (2005). Anonymity in giving in a natural context – A field experiment in 30 churches. *Journal of Public Economics, 89*(11–12), 2301–2323. doi:10.1016/j.jpubeco.2004.11.002

Spraggon, J. M., Sobarzo, L. A. V., & Stranlund, J. K. (2015). A note on stochastic public revelation and voluntary contributions to public goods. *Economics Letters, 126*, 144–146. doi:10.1016/j.econlet.2014.12.009

Velez, M. A., Stranlund, J. K., & Murphy, J. J. (2009). What motivates common pool resource users? Experimental evidence from the field. *Journal of Economic Behavior & Organization*, *70*, 485–497. doi:10.1016/j.jebo.2008.02.008

Yamagishi, T. (1986). The provision of a sanctioning system as a public good. *Journal of Personality and Social Psychology*, *51*(1), 110–116. doi:10.1037/0022-3514.51.1.110

Zafar, B. (2011). An experimental investigation of why individuals conform. *European Economic Review*, *55*(6), 774–798.

# PART III
# REPLICATION AS A BASIS FOR EXTENSION

# ON REPLICATION AND PERTURBATION OF THE MCKELVEY AND PALFREY CENTIPEDE GAME EXPERIMENT

James C. Cox and Duncan James

## ABSTRACT

*This study first replicates, then perturbs, the centipede game as implemented by McKelvey and Palfrey (1992). It is thus both a replication study and an original research study. We use controlled laboratory experiments, with computer interfaces for each treatment, anonymous round-robin matching among the subjects across rounds, multiple (10) rounds within each treatment, and incremental changes between adjacent treatments allowing for an assessment of effects at the margin of different game configurations. We find unraveling to the subgame perfect equilibrium somewhat faster than did McKelvey and Palfrey (1992), when using their exact design. Perturbations to that design show that setting non-taker payoffs to zero induces earlier unraveling, as does the use of higher stakes (as in Murphy, Rapoport, and Parco (2006), and Rapoport, Stein, Parco, and Nicholas (2003), respectively). Other, subsequent perturbations show: that there is at most a subtle effect*

**Replication in Experimental Economics**
**Research in Experimental Economics, Volume 18, 53–94**

**ISSN: 0193-2306/doi:10.1108/S0193-230620150000018003**

*associated with using a 10-second timer with a default move, relative to untimed active moves; and that clock format versus tree format has a minimal effect in common information, unchanging payoff-parameterization environments. We verify the robustness of some key past findings in real-time games. We also explore in a common information environment, the effect of design features previously used in independent private values settings; here we find new evidence that features which might modulate information acquisition and/or processing in an independent private values setting may not restrict behavior in a common information setting.*

**Keywords:** Games; auctions; experiments; information

**JEL classifications:** C70; C72; C92; D02; D44; D80

## INTRODUCTION

McKelvey and Palfrey (1992) present one of the most thought-provoking results in experimental testing of game theory. The centipede game they test was purposely built as a paradox (Rosenthal, 1981), pitting geometric accumulation, naïve optimism, and noise (and maybe altruism) against the gravitational pull of backward induction. Brought into the lab, it did not disappoint. Backward induction – or more formally subgame perfection – failed in the face of motley opposition, at least over 9 or 10 repetitions of the game with round-robin matching.

McKelvey and Palfrey's (1992) centipede game experiments began with a total amount of money, \$0.50 that was divided into a large share, \$0.40, and a small share, \$0.10. Both of these amounts were doubled each time a player chose the action "Pass" rather than the action "Take". The player who chose Take in any round got the larger share and the other got the smaller share. McKelvey and Palfrey experimented with two round (or four move) and three round (or six move) centipede games. They also experimented with a four move game with quadruple the above payoffs. Five of the experimental sessions included 20 subjects and the other two sessions included 18 subjects. The subjects in a session were divided into two equal-sized groups (of 10 or 9) and each subject in one group played every subject in the other group once. Subjects in the experiments were undergraduate students at Pasadena Community College and the California Institute of Technology.

The standard game theoretic prediction of outcomes for selfish players who do not make errors (and believe that others are selfish and do not make errors) is that the player with the option to choose Take on the first move of the game will do so. Behavior in the experiments diverged widely from this prediction. Only 37 out of 662 of McKelvey and Palfrey's (1992) experimental centipede games ended with a take on the first move, 23 of the games ended with both players passing at every move, and 602 games ended with a take at an interior move. The authors proposed an alternative model of altruistic players to explain the data.

Subsequent research on centipede games has continued along several lines of inquiry. Models of noise in perception (Zauner, 1999) or response (Fey, McKelvey, & Palfrey, 1996; McKelvey & Palfrey, 1998) is one line of inquiry. Attempted perturbation of subject pools is another (e.g., Palacios-Huerta & Volij, 2009). Both of the preceding lines of inquiry focus on the characteristics of individual agents. Another approach is to modify the game itself – in terms of presentation, move structure, payoff scale, and other features. Fey et al.'s (1996) use of a centipede game in which the sum of payoffs across players was constant (rather than geometrically increasing) was the first such modification. Nagel and Tang (1998); Rapoport, Stein, Parco, and Nicholas (2003); Murphy, Rapoport, and Parco (2006); and Cox and James (2012) also pursued variations of this approach of altering features of the game in more or less fundamental ways. All such approaches of altering the game have in common that they seek to find structures and parametrizations, exogenous to the agent but controllable by the experimenter, which modulate behavior in the centipede game.

In the present paper, we report attempts to replicate McKelvey and Palfrey's (1992) findings using (a) their baseline nominal payoffs and (b) the 2014 current (CPI-adjusted) value of their baseline payoffs. We also select three alterations in game form found in previous work to significantly modulate centipede game behavior and apply them to the original McKelvey and Palfrey (1992) design. We thus bring subsequent discoveries back to the original experimental setting and allow for a mutual robustness test, in addition to what is intended to be an attempt at replication of the original McKelvey and Palfrey (hereafter MP) findings for play of the game.

In addition to an attempted replication of MP, our perturbations examine the effects of changing features of the MP game, one at a time. Each perturbation is interesting in its own right, as it allows the parsing of marginal contributions to unraveling, and a comparison with earlier research centered on MP. However, we can then also branch out into comparisons with different lines of research. Notably, Cox and James (2012) point out that Dutch auctions and centipede games have a similar structure. The sequence of perturbations in this present replication exercise also then delivers us to parameterizations which can revisit the "clock format versus tree format" question about institutional structure from Cox and James (2012) – except in a very different setting: the MP environment. This then sets up mutual robustness checks: Does the MP design generate different results for different formats? Does the Cox and James (2012) finding of different results across different formats persist in a common information, geometric accumulation environment?

The sequence of perturbations is:

(A) *Payoff scale.* Switching between the MP nominal and MP current (CPI-adjusted) payoffs lets us ask whether variation of the size of the aggregate payoff modulates behavior across our experiments with the MP design, as it did for a different (non-MP) centipede game design used by Rapoport et al. (2003).
(B) *Zero lower payoff.* By setting to zero the payoff of the player who does not choose Take we ask whether this alteration changes behavior, relative to the unaltered MP design with positive payoffs for the non-taker. Results from Murphy et al. (2006) are suggestive that it might, but actually checking this in the MP design – contrasted against the original MP setting of positive payoffs to the non-taker – is the only way to find out directly.
(C) *Timed moves.* Cox and James (2012) used a 10-second timer (with Pass/"not purchase" as the default option) throughout all their experiments. They did so in order to standardize the treatment of time in a research program encompassing both the Dutch auction and the centipede game. But the use of a timer in the centipede game is an innovation worth assessing in its own right, to see what effect it may have on its own. We address this question by introducing a 10-second timer, and Pass as the default action, to the MP centipede game (with zero payoff to non-takers) in tree format. This treatment can then be compared to the tree format MP centipede game (with zero payoffs to non-takers) without a timer, and with all moves active.
(D) *Clock format.* The choice of tree or clock institutional format for the centipede game has been found to change outcomes fundamentally when there is *private* information about payoffs. Cox and James (2012) report that (10-second) clock format centipede games with independent private values (IPV) produce outcomes that converge more rapidly to the theoretical prediction of unraveling to a choice of Take at the first opportunity than otherwise identical (e.g., also using timed

nodes) centipede games in tree format. This suggests the question of whether there will be faster unraveling with a clock format and *common* information about payoffs. We address this question by finally introducing a clock format in the MP centipede game. We then compare results across clock and tree formats, holding constant across both formats: the use of the 10-second timer, use of Pass as the default action, zero payoff to non-takers, and otherwise the original MP parameterization, replete with common information.

A preview of our results is as follows. We first experiment with the MP four-decision-node, geometric accumulation, common information centipede game with positive payoffs for the player who does not choose Take. Our baseline replication results from experimenting with this game, with stakes inflation-adjusted so as to be comparable to current-dollar value of the stakes in the MP "low treatment", are at the lower boundary of the results from MP. That is, our data show more unraveling than most of the corresponding treatments in their paper (although our data corresponds fairly closely to one of their Pasadena Community College sessions, PCC-2).

From that point we begin our sequence of perturbations. We implement the game with the same nominal (unadjusted) payoffs used by MP; at these lower *current* dollar stakes, unraveling takes place more slowly. Our finding of faster unraveling at higher stakes is consistent with findings in MP, and with those in Rapoport et al. (2003).

We also perturb the game from MP by setting equal to zero the payoffs to any player not choosing Take. We do this both as a robustness check on the results in MP, and as a way of connecting more closely their original results with those in recent centipede game experiments in Cox and James (2012). We find that this design change induces earlier choices of Take compared to otherwise comparable parameterizations of the original MP centipede game. Of note, the effect of having one player receive zero if the other player chooses Take appears to be similar to increasing the total size of the "pot" (positive lower plus higher payoff). This also further clarifies results from Murphy et al. (2006) on changing the proportions according to which total payoffs at a node were allocated across the one "winner" and the $n-1$ "losers".

We find that the imposition of timed moves and a default action of Pass does not have a large effect, though to the extent there is one it is in the direction of earlier choice of Take when timed moves and default Pass are included in the design.

Finally we find that in a common information, unchanging-payoff-structure-across-rounds setting, clock format versus tree format does not make much difference to behavior (unlike in an IPV environment reported by Cox & James, 2012). This suggests that the MP results are robust to the tree versus clock format change provided that the information at each payoff node never changes over the course of the experiment, and that each player has had an opportunity to view all the other player's payoffs. Conversely, taken with the Cox and James (2012) results this suggests that tree versus clock institutional format has a significant effect on observed unraveling in a changing, private information (on monetary payoffs) environment but not in a static, common information environment.

## REPLICATION AND PERTURBATION OF THE MP EXPERIMENT

### *Experimental Design and Protocol*

The experiment reported here was run in the laboratory of the Experimental Economics Center (E*x*CEN) at Georgia State University in Atlanta in 2014. The experiment was computerized. Subjects participated in groups of 40 (with the exception of one group of 20, for purposes of replicating MP). Subjects were randomly selected from the database of undergraduate students who had previously volunteered to participate in experiments at E*x*CEN.

The centipede game from MP, with doubled (CPI-inflation-adjusted) payoffs, appears in Fig. 1.[1]

It is a game of common information; each player can see all payoffs of both players. The summed payoffs (across the two players) double from one node to the next, left to right. The manner in which the split of this payoff "pot" alternates from one decision node to the next sets up a stark contrast at the final decision node – which is predicted to influence choice at the prior decision node, and so on, back to the first decision node.

In the original MP study, round-robin matching was used with 18 or 20 subjects, guaranteeing to the subjects that they would never face the same opponent twice, and thus that no reputation formation could occur. MP ran subjects through only 10 rounds. We use 10 rounds in our inflation-adjusted replication. We also use round-robin matching with 20 subjects in our replication treatment.

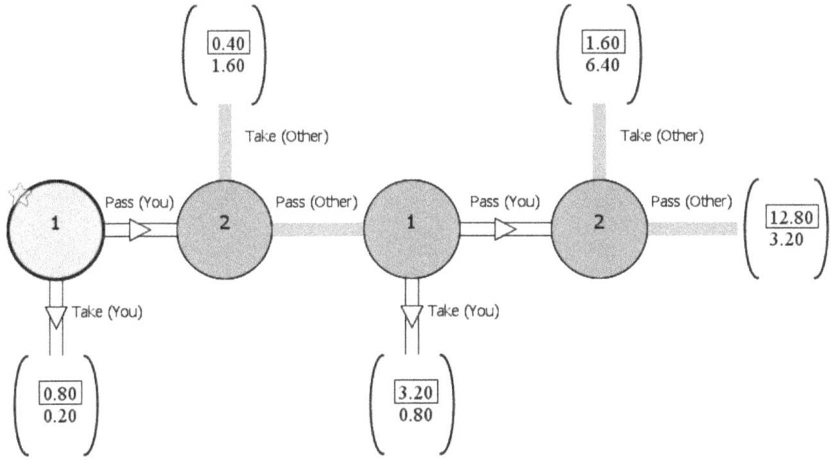


*Fig. 1.* Game Tree for the T1 Inflation-Adjusted Treatment.

In our perturbation treatments, we use larger pools of subjects so as to allow such matching for a greater number of rounds. Specifically, we used groups of 40 for experiments consisting each of three segments of 10 rounds, so that no subject ever faced another more than once over 30 rounds. In these treatments we run the subjects through an ABA or BAB sequence of three 10-round segments. In these perturbation treatments the first 10-round segment should still be comparable to the MP data because the existence of subsequent BA or AB rounds in an ABA or BAB treatment was not announced before completion of the first 10 rounds.

Our attempt at replication of the MP experiment takes place in the treatment we call the T1-Inflation-Adjusted treatment. We call the treatment with their nominal (lower real) stakes T1. The payoffs in T1 are one-half the payoffs shown in Fig. 1. The T2 treatment is the same as T1 except the payoffs to the non-taker are set to zero. The T3 treatment is the same as T2 except there is a 10-second take opportunity which renders the Pass action passive (it occurs after 10 seconds, and not before, and not due to mouse clicks). Treatment T4 is like T3 except that the tree institutional format is replaced by a digital price clock format familiar from the Dutch auction, as illustrated in Fig. 2. Note that both T3 and T4 have the active Take, passive Pass after 10 seconds, temporal structure just mentioned; tree versus clock refers to visual format, as in the difference between Figs. 1 and 2, and not any difference in temporal structure – *there isn't one.*

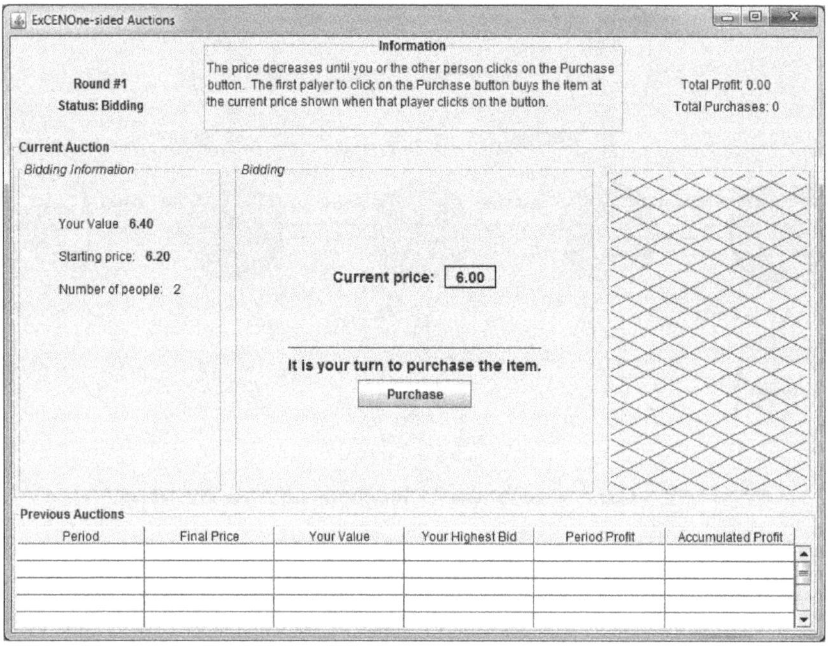


*Fig. 2.* Digital Price Clock Format for T4.

The treatments with 10-second take opportunity (T3 and T4) are used to produce data assessing whether some findings from the experiment on centipede games in Cox and James (2012) extend from an IPV environment to a common information environment, and for adding additional robustness checks on the MP experiment. We summarize our treatments in Table 1. We compare our design to MP in Table 2.

In all experiment sessions, subjects were paid the sum of their payoffs from all rounds, as in the MP experiment.

Subjects read the instructions on their computer monitors and were invited to raise their hands if they had questions. When a hand was raised, an experimenter approached the subject's privacy-screened computer station; the question was subsequently asked, and answered, in private. After all subjects had clicked on the "Finished Reading" button on their monitors, an experimenter projected and read aloud a summary of the subject instructions. Subjects were invited to raise their hands during this reading if they had questions; again, any questions were asked, and answered, in

***Table 1.*** Treatments in Current Experiment.

| Treatment | Taker Payoffs | Non-Taker Payoffs | Format | Timer |
|---|---|---|---|---|
| T1 Infl. Adjusted | Twice MP "low" | Twice MP "low" | Tree | No |
| T1 | MP "low" | MP "low" | Tree | No |
| T2 | MP "low" | zero | Tree | No |
| T3 | MP "low" | zero | Tree | Yes |
| T4 | MP "low" | zero | Clock | Yes |

***Table 2.*** Treatments in MP and Current Experiment.

| Design Attribute | McKelvey and Palfrey (1992) | Current Study | Cox and James (2012) |
|---|---|---|---|
| Information environment | Common | Common | IPV |
| Payoff to player other than the player choosing Take | 20% of aggregate payoff pool at that node | Zero or 20% of aggregate payoff pool at that node (treatment variable) | Zero |
| Format | Tree | Clock or tree (treatment variable) | Clock or tree (treatment variable) |
| Payoff scale | Higher or lower (treatment variable) | Higher or lower (treatment variable) | Not a treatment variable (however, lower average subject payout than this study) |
| Move order | Alternating | Alternating | Alternating or concurrent (treatment variable) |
| Timer | Untimed | Untimed or timer with Pass as a default move after 10 seconds with no take (treatment variable) | Timer with Pass as a default move after 10 seconds with no take |

private. Subject instructions and projection files for all treatments are available at http://excen.gsu.edu/jccox/instructions/

## *Results from Replication Attempt for the MP Experiment*

Fig. 3 reports the average take node (across subject pairs in a round) in the MP experiment and in our replication; standard deviation either side of the

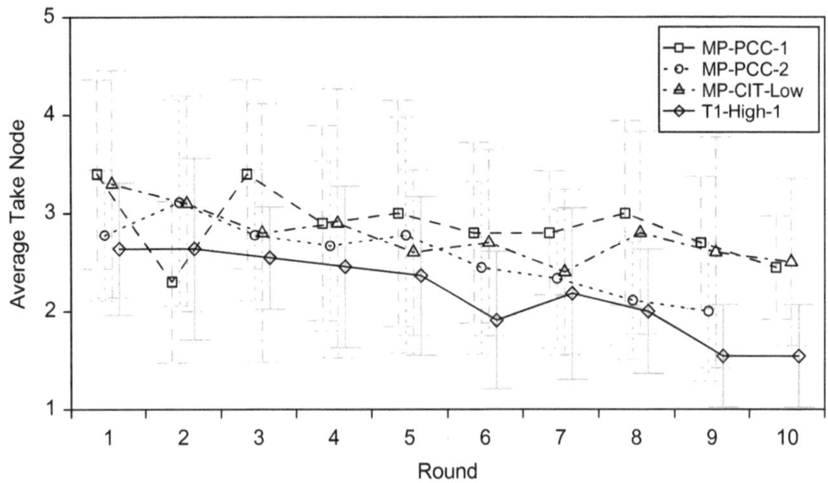


*Fig. 3.* Direct Comparison to MP Data.

average is represented by "whiskers"; PCC refers to sessions run at Pasadena Community College; CIT refers to sessions run at Cal Tech. Our first finding is that the MP result that play does not completely unravel to a take at the first node can be replicated with the same four decision node/five payoff node game they used and the same real payoffs they used. Our T1-Inflation-Adjusted data are closest to data from MP's Session 2, run at Pasadena Community College (MP PCC-2 in Fig. 3). But we find more unraveling than they found overall; our average take node is 1.55 in rounds 9 and 10 whereas their round 9 and 10 average take nodes were 2.5 or higher in all treatments except PCC-2. Using the game as an independent unit of observation, our replication is statistically different from all of the MP treatments in terms of central tendency; our results exhibit earlier choice of Take (the highest $p$-value being 0.0049 in pairwise rank-sum tests).[2] Note, however, that the first period of our replication does not differ from any of the MP treatments at the 5% level (with the lowest $p$-value being 0.0604 for the pairwise comparison with MP-PCC-1).

### *Results from Perturbation: Role of Payoff to the Non-Taker*

Perturbation of the original MP design allows an assessment of the role of positive payoffs to non-takers (vs. zero, as would be natural in the centipede game's close relative, the Dutch auction). T1 employs the MP design with their nominal payoffs (there is no inflation adjustment). T2 is the same as T1 except payoffs to non-takers are zero.

The first panel of Fig. 4 plots average take node data for the session with 10 rounds of T1, followed by 10 rounds of T2, followed by 10 rounds of T1: the T1-T2-T1 session. In this session we observe strong unraveling in

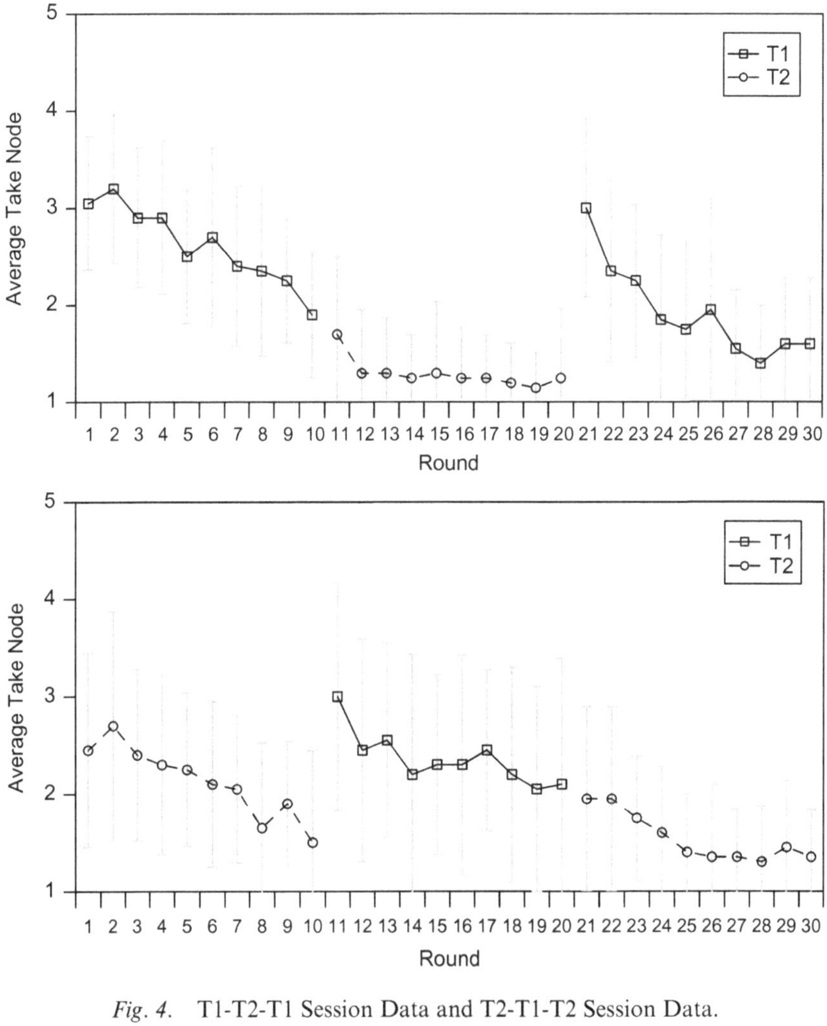


*Fig. 4.* T1-T2-T1 Session Data and T2-T1-T2 Session Data.

treatment T2 beginning in round 2 for this treatment (round 12 in the session).

The second panel of Fig. 4 plots data for the session with 10 rounds of treatment T2, followed by 10 rounds of treatment T1, followed by 10 rounds of treatment T2: the T2-T1-T2 session. There is strong unraveling in the second implementation of T2 (see rounds 25–30 in the second panel of Fig. 4).

If we pool across all T1 games and T2 games (treating them as independent observations), we find that T2 supports a choice of Take which is 0.645 nodes earlier than T1, on average. Based on this pooled data, there is statistically different central tendency between T1 and T2 ($p < 0.001$ via a rank-sum test). We also find that this holds for the first period (in which all games presumably *are* independent) as well (with $p$-value of 0.049).

The data plotted in Fig. 4 can also be evaluated statistically in a random effects regression in which the dependent variable is node at which Take occurs for a game/pair and the explanatory variables are: time trend, dummy variable for Treatment 1, time trend × dummy variable for Treatment 1, dummy variable for 2-1-2 session, time trend × dummy variable for 2-1-2 session × dummy variable for Treatment 1.

$$\text{Take node}_{it} = a + B_1 \times t + B_2 \times T1 + B_3 \times t \times T1 + B_4 \times (212 \text{ session}) + B_5 \times (212 \text{ session}) \times t \times T1 + \varepsilon_{it}$$

With 1,200 observations and random effects at the game (pair) level this yields the coefficient estimates in Table 3.

Clearly, T1 generates later choices of Take than T2. This is the case even when subjects have had 10 rounds of T2 in which to move toward choosing

***Table 3.*** Effects on Unraveling of Zero Non-Taker Payoffs.

| | Coefficient | $p$-value |
|---|---|---|
| Intercept | 1.824 | < 0.001 |
| Time trend | −0.034 | 0.009 |
| Intercept dummy for T1 | 1.095 | 0.003 |
| Slope dummy for T1 | −0.007 | 0.687 |
| Intercept dummy for T2-T1-T2 session | 0.545 | 0.020 |
| Slope dummy for T1 in T2-T1-T2 session | −0.029 | 0.159 |
| Std. dev. of the random effect | 2.773 | |
| Log-likelihood | 2850.4 | |

Take at the first node, as in the T2-T1-T2 session; here we see the trend toward taking at the first node, as established in the first T2 segment, is broken by the introduction of T1.

## *Results from Perturbation: Role of Stakes*

Data from our T1 Inflation-Adjusted treatment (presented along with comparable MP data in Fig. 3) can alternatively be compared to the first segment T1 data from Fig. 4, in order to get an idea of the "marginal" effect of stake-level in our study. T1 and T1 Inflation-Adjusted plotted together yields Fig. 5.

If we pool across the first ten T1 games (of the T1-T2-T1 session) and all of the T1 Inflation-Adjusted games (treating them as independent observations), we find that there is a statistically different central tendency, with an earlier choice of Take in T1 Inflation Adjusted relative to T1 (with a p-value of 0.0242 via a rank-sum test). On average, T1 Inflation-Adjusted yields a choice of Take 0.41 nodes earlier than T1. However, if one restricts attention to only the first round, then with this smaller sample one cannot reject that central tendencies are the same ($p$-value of 0.171).

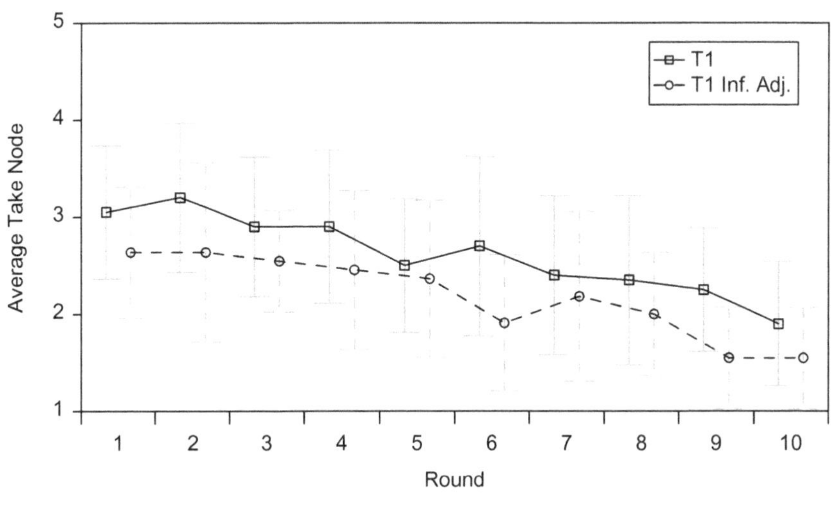


*Fig. 5.* T1 and T1 Inflation-Adjusted Data.

One interesting conclusion that might be drawn, from the results in this section and those in the section on payoffs to non-takers, is that zero payoffs to non-takers (as in Cox & James, 2012) yields earlier unraveling – similar, if not greater than the effect of higher stakes (as in MP, or Rapoport et al., 2003).

### *Results from Perturbation: Role of Timed Decision Opportunity*

Some light on the role of the timer (or its lack) might be shed by comparing the first segments of the T3-T4-T3 and T2-T1-T2 sessions. Comparing the early T2 and T3 segments in cross section yields the following.

Visually, the average node at which Take is chosen is barely distinguishable across treatments. The first segments T2 and T3, shown in Fig. 6, are compared directly, since the subjects at that point have not been exposed to T1 or T4, respectively. If we then pool over 10 rounds for each treatment (treating each game within each treatment as an independent observation), we do not find a statistically different central tendency in take node at the 5% level (with a $p$-value of 0.0721 via a rank-sum test).[3] It should also be noted that whatever the statistical standard, the economic magnitude of any difference between T2 and T3

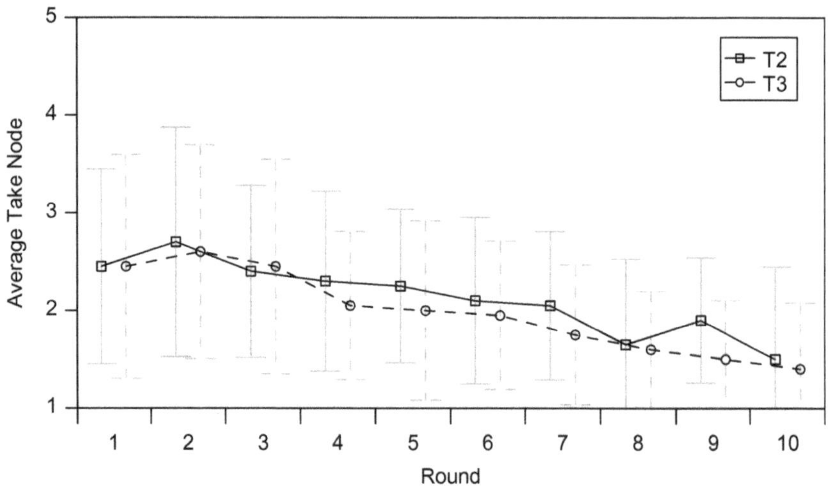


*Fig. 6.* T2 and T3 First Segment Data.

would appear to be small; on average, T3 (timer, and pass as default move) games end 0.155 nodes earlier than T2 (untimed, all moves actively chosen) games. In this setting, at least, having a 10-second timer and Pass as the default action does not appear to affect behavior much relative to untimed, actively chosen actions.

### *Results from Perturbation: Role of Clock Format versus Tree Format*

The MP design can be adapted to incorporate some of the features present in the recent centipede game experiment in Cox and James (2012). Suppose one were to start by maintaining the common information, stationary-across-rounds payoff structure of MP, rather than an IPV (redrawn each round) environment of Cox and James (2012). That could be augmented with a 10-second clock and passive Pass actions (forced by the 10-second timer), and zero payoff to non-takers; doing so yields our treatment T3.[4] Treatment T3 then translated visually to a Dutch-auction-style clock format yields treatment T4. These treatments introduce changes at the margin to the MP complete information game, not to the Cox and James (2012) private information games. These treatments thus do not allow replication of the results from Cox and James (2012), but rather a robustness test in a different environment.

We find the following for T3 and T4. First, note that the second set of T3 rounds in the first panel of Fig. 7 includes observations of almost complete unraveling. In particular, rounds 27 and 29 in the first panel of Fig. 7 show 19 choices of Take at the first node (out of 20 subject pairs) and one choice of Take at the second node. (These rounds correspond to rounds 7 and 9 of the repeat/order control segment for the T3 treatment.) The tree version supports takes which are slightly earlier than the clock version - the average node at which Take is chosen for T3 is 1.56, while for T4 it is 1.68. This difference of 0.12 nodes on average is economically small. For scale, consider that Cox and James (2012) reported a value of 0.78 nodes as the average difference between clock format and tree format (in node/price at which Take was chosen). There, the tree format supported later rather than earlier choice of Take.

If we pool, by treatment, across these two sessions (treating each game as an independent observation), we find that there is a statistically different central tendency, with an earlier adoption of Take in T3 relative to T4 (with a *p*-value of 0.0036 via a rank-sum test). However, if one restricts

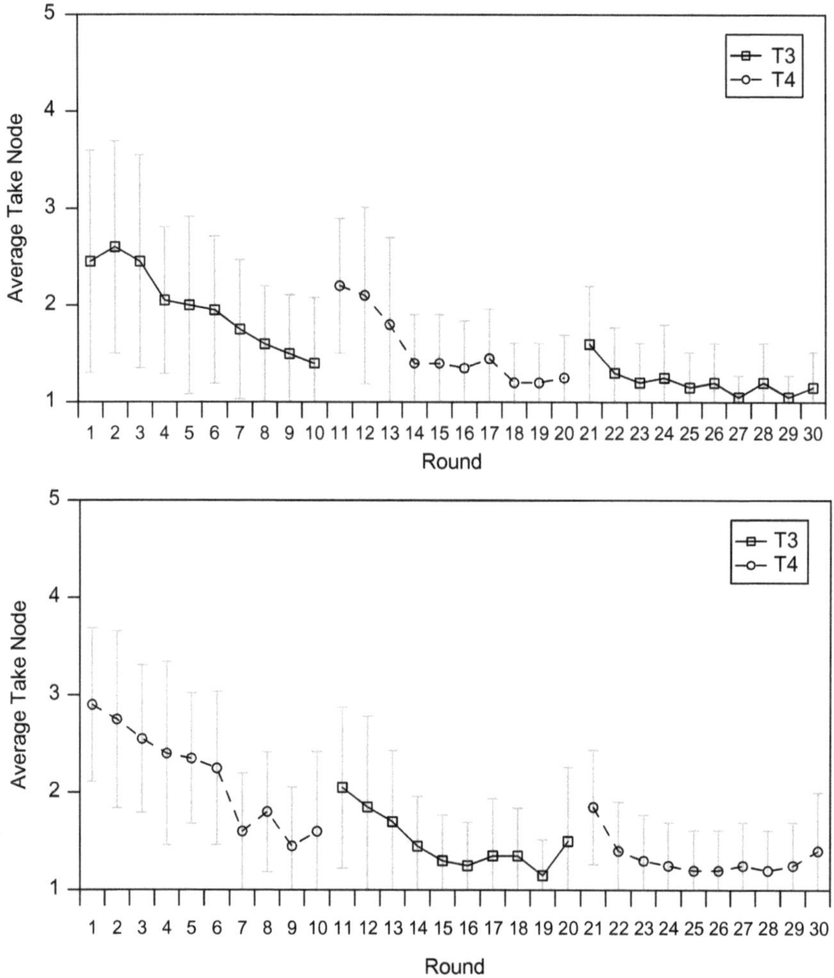


*Fig. 7.* T3-T4-T3 Data and T4-T3-T4 Data.

attention to only the first round, then one cannot reject that central tendencies are the same ($p$-value of 0.157).

What do random effects regression results suggest? Coefficient estimates are reported in Table 4 from estimation of a random effects (at the game/pair level) model with 1,200 observations, in which the dependent variable is node at which Take occurs for a game/pair, and the explanatory variables

***Table 4.*** Effects of Clock and Tree Institutional Formats.

| | Coefficient | *p*-value |
|---|---|---|
| Intercept | 2.280 | < 0.001 |
| Time trend | –0.048 | < 0.001 |
| Intercept dummy for T3 | –0.079 | 0.366 |
| Slope dummy for T3 | 0.008 | 0.055 |
| Intercept dummy for T4-T3-T4 session | 0.221 | < 0.001 |
| Slope dummy for T3 in T4-T3-T4 session | –0.020 | < 0.001 |
| Std. dev. of the random effect | 0.658 | |
| Log-likelihood | 2651.9 | |

are: time trend, dummy variable for Treatment 3, time trend × dummy variable for Treatment 3, dummy variable for 4-3-4 session, time trend × dummy variable for 4-3-4 session × dummy variable for Treatment 3.

$$\text{Take node}_{it} = a + B_1 \times t + B_2 \times T3 + B_3 \times t \times T3 + B_4 \times (434 \text{ session}) + B_5 \times (434 \text{ session}) \times t \times T3 + \varepsilon_{it}$$

Such difference as there is between treatment T3 and T4 data, while statistically significant in the pooled nonparametric test, is not statistically significant in the random effects regression. Moreover, the coefficient on T3 is, in keeping with the earlier comparison of raw averages, small in absolute and economic magnitude.

These results suggest that for repeated payoff structures with *common* information, tree vs. clock format may not make a difference. But this also illustrates the potentially important interaction between environment and institution, for in an IPV environment institutional format most certainly does make a difference, both statistically and in economic magnitude: see Dorsey and Razzolini (2003) and Turocy, Watson, and Battalio (2007) for auction results; and see Cox and James (2012) for both IPV centipede game and IPV auction results.

## CONCLUSIONS

In experiments that allow for reproducing earlier findings by other researchers we find the following: Our inflation-adjusted replication treatment that most closely corresponds to the McKelvey and Palfrey (1992) "low" treatment produces more unraveling than their experiment (Fig. 3). We do replicate earlier results for the effects of changing (nominal) monetary stakes: higher stakes elicit more unraveling, as in MP and Rapoport et al. (2003). We also find that earlier choice of Take is economically and statistically significantly promoted by zero payoffs to the non-taker in the MP environment, echoing Murphy et al. (2006) in a different common information environment.

In experiments branching out from replication, to test new hypotheses, we generate fresh findings which are interesting in their own right. A robustness check on the impact of the introduction of timed nodes with Pass as the default action to the MP design (with zero payoffs to non-takers) finds that the addition of this design feature has a small magnitude effect (in favor of earlier choice of Take when the timed-nodes-and-Pass-as-default-action design feature is in place) of questionable statistical significance. Next, introduction of a clock visual format to the MP common information centipede game (with timed-nodes and default of Pass, and zero payoffs to non-takers, already in place from prior perturbations) allows assessment of the effect of using that format rather than a tree format. In the MP common information, unchanging payoffs setting, the clock format supports choice of Take 0.12 nodes later than the tree format, in contrast to Cox and James (2012) who found the clock format producing choice of Take 0.78 nodes earlier (in an IPV centipede setting). Our interpretation of this dovetails with our interpretation, stated in Cox and James (2012), as to the likely mechanism by which institutional format might affect behavior. That is, we conjecture that institutional format interferes with information acquisition and processing. This would have a noticeable effect in an IPV environment, where (private information) parameters are redrawn every round; and this is what is observed in Cox and James. Conversely, in the MP environment, the (common information) parameters are the same every round, and are disclosed prior to the start of even the first round, by means of the subject instructions. In a sense, in the MP environment by the time play starts there is no more information to acquire (about the parameterization of the game, at least).

Overall, we have generated a verified and expanded set of empirical results that future researchers can reference and use when designing and analyzing experiments in real-time game theory.

# NOTES

1. The payoffs at corresponding nodes in the MP "low" treatment are one-half of those in Fig. 1. The MP paper was submitted to *Econometrica* in May 1990. We guess that their experiment was run in 1989 or earlier. Our experiment was run in May and June 2014. Using the CPI, one would calculate that $1 in 1989 has the same purchasing power as $1.92 in 2014. Doubling the MP nominal payoffs provides a reasonably good inflation adjustment that also provides nice numbers for use in the game tree.

2. The argument can be made that each game is not fully independent of all other games. However, we do employ round-robin matching, and the subjects are informed of this, so any violation of independence should not come from attempts at reputation formation.

3. If pooling across all T2 and T3 games (including those after exposure to T1 and T4 respectively), the difference in central tendency between T2 and T3 is also insignificant at the 5% level (with a $p$-value of 0.062).

4. Use of the 10-second timer for both clock (T4) and tree (T3) formats (the nomenclature refers to their visual appearance, not their identical temporal structure) is for the purpose of not introducing any avoidable differences in the clock versus tree comparison, relative to that which has already been explored (Cox & James, 2012). The original use of such a timed move structure in the centipede game, in Cox and James (2012), was for the purpose of standardizing temporal structure across different institutions, and doing so in such a manner as to implement that which is usual in the Dutch auction (passive pass, active take, on some time interval). Whether or not this type of timed move imparts some kind of framing effect or cue in and of itself is one thing we try to assess in this present study (wherein we compare the first segments of T2 and T3, T2 being untimed, while T3 has the timer just described). However, the comparisons between clock and tree format, that is, between T3 and T4, are between treatments which *both* employ the 10-second, passive pass, active take, timer feature.

# REFERENCES

Cox, J. C., & James, D. (2012). Clocks and trees: Isomorphic Dutch auctions and centipede games. *Econometrica, 80*, 883–903.

Dorsey, R., & Razzolini, L. (2003). Explaining overbidding in first price auctions using controlled lotteries. *Experimental Economics, 6*, 123–140.

Fey, M., McKelvey, R., & Palfrey, T. R. (1996). An experimental study of constant-sum centipede games. *International Journal of Game Theory, 25*, 269–287.

McKelvey, R., & Palfrey, T. R. (1992). An experimental study of the centipede game. *Econometrica, 60*, 803–836.

McKelvey, R., & Palfrey, T. R. (1998). Quantal response equilibria for extensive form games. *Experimental Economics, 1*, 9–41.

Murphy, R. O., Rapoport, A., & Parco, J. E. (2006). Breakdown of cooperation in iterative real-time trust dilemmas. *Experimental Economics, 9*, 147–166.

Nagel, R., & Tang, F. F. (1998). Experimental results on the centipede game in normal form: An investigation on learning. *Journal of Mathematical Psychology*, *42*, 356–384.

Palacios-Huerta, I., & Volij, O. (2009). Field centipedes. *American Economic Review*, *99*, 1619–1635.

Rapoport, A., Stein, W E., Parco, J E., & Nicholas, T E. (2003). Equilibrium play and adaptive learning in a three person centipede game. *Games and Economic Behavior*, *43*, 239–265.

Rosenthal, R. (1981). Games of perfect information, predatory pricing, and the chain-store paradox. *Journal of Economic Theory*, *25*, 92–100.

Turocy, T L., Watson, E., & Battalio, R. C. (2007). Framing the first-price auction. *Experimental Economics*, *10*, 37–51.

Zauner, K G. (1999). A payoff uncertainty explanation of results in experimental centipede games. *Games and Economic Behavior*, *26*, 157–185.

# APPENDIX

This appendix contains the instructions for T1-Inflation-Adjusted, T3, and T4. The instructions for T1 and T2 are omitted to save space, as they make only minor adjustments relative to T1-Inflation-Adjusted, and T3, respectively. Instructions for all treatments are available at http://jccox/instructions/

## *INSTRUCTIONS (T1-INFLATION-ADJUSTED)*

This is an experiment in the economics of decision-making.

During the experiment you will be participating in a decision procedure in which the participants **take turns** in **choosing between two actions**. Those actions are **"Take" or "Pass"**.

If the participant whose turn it is to choose were to select the action **"Take"**, that would **end the decision procedure**. On the other hand, if the participant whose turn it is to choose were to decide to **"Pass"**, then the **proceedings** would **continue**.

You will be **randomly selected to be either Player 1 or Player 2**. You will **keep your role** as Player 1 or Player 2 **throughout the experiment**.

Player 1 will always be the first person to choose Pass or Take at the first node. If Player 1 chooses to Pass, Player 2 will get the next turn to Take or Pass.

*LOOKING AT DECISION OPPORTUNITIES FOR PLAYER 1 AT NODE 1*

If you are randomly selected to be Player 1 then the below tree diagram shows your available actions and payoffs. The label (1 or 2) at each node indicates who will make a decision at that node.

Player 1's payoff at a given branch of the tree is the top number; it is outlined by a rectangle in the below tree diagram. The other number, not in the rectangle, is the other person's payoff.

In order for **Player 1** to **Take at** the **first decision node**, he/she needs to click on the Take arrow; in that case **Player 1 is paid $0.80 and Player 2 is paid $0.20**.

In order for **Player 1** to **Pass at** the **first decision node**, he/she needs to click on the Pass arrow; in that case there are **no money payoffs so far and Player 2 gets the next turn** to Take or Pass.

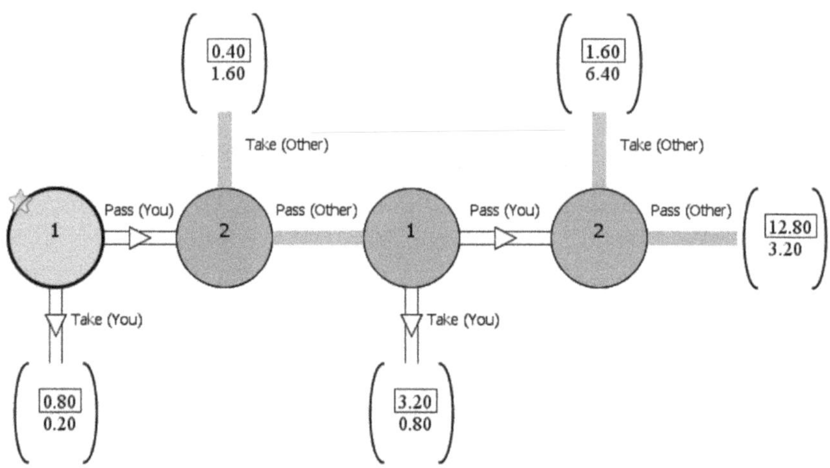

## *LOOKING AT DECISION OPPORTUNITIES FOR PLAYER 2 AT NODE 2*

If you are randomly selected to be Player 2 then the below tree diagram shows your available actions and payoffs. The label (1 or 2) at each node indicates who will make a decision at that node.

Player 2's payoff at a given branch of the tree is the bottom number; it is in a rectangle in the below tree diagram. The other number, not in the rectangle, is the other person's payoff.

In order for **Player 2** to **Take at** the **second decision node**, he/she needs to click on the Take arrow; in that case **Player 2 is paid $1.60 and Player 1 is paid $0.40.**

In order for **Player 2** to **Pass at** the **second decision node**, he/she needs to click on the Pass arrow; in that case there are **no money payoffs so far and Player 1 gets the next turn** to Take or Pass.

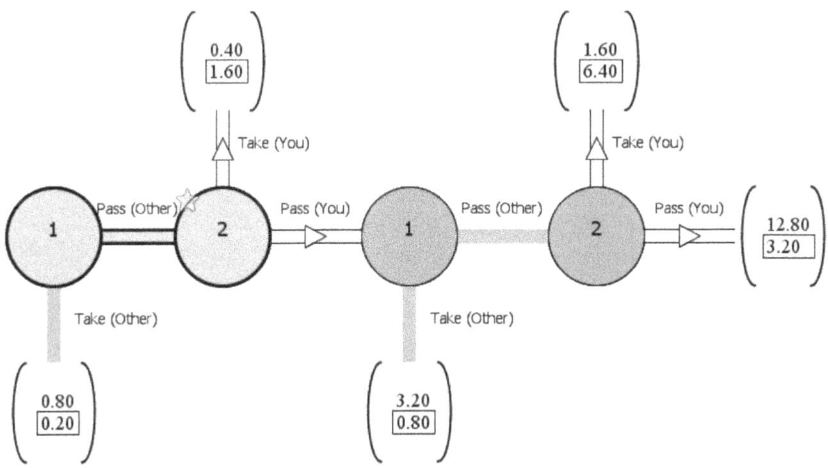

*LOOKING AT DECISION OPPORTUNITIES FOR PLAYER 1 AT NODE 3*

If **Player 1** clicks on **Take at the third decision node** then **Player 1 is paid $3.20 and Player 2 is paid $0.80.**

If **Player 1** clicks on **Pass at the third decision node then there are no money payoffs so far and Player 2 gets the next turn** to Take or Pass.

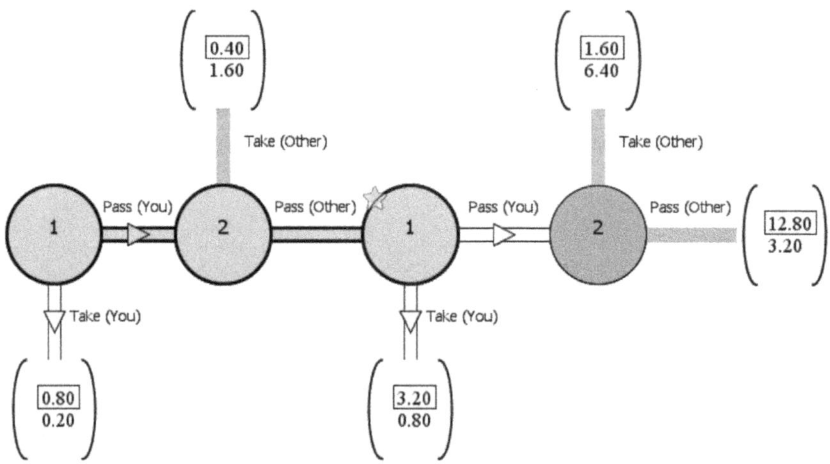

## *LOOKING AT DECISION OPPORTUNITIES FOR PLAYER 2 AT NODE 4*

If **Player 2** clicks on **Take at the fourth decision node** then **Player 2 is paid $6.40 and Player 1 is paid $1.60.**

If **Player 2** clicks on **Pass at** the **fourth decision node** then the decision procedure ends with the payoffs at the end of the tree: **Player 2 gets paid $3.20 and Player 1 gets paid $12.80**.

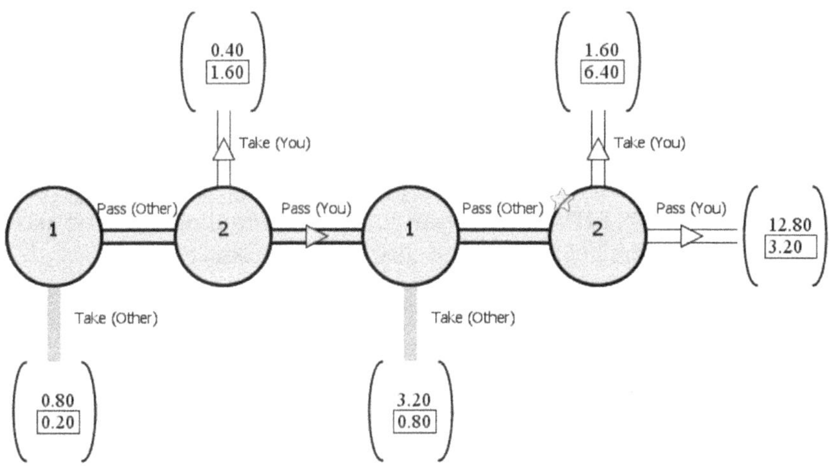

*MANY ROUNDS BUT DIFFERENT PLAYERS*

You will **participate** in the decision procedure **many times**; each participation repetition is called a "round".

You will be **randomly selected as either Player 1 or Player 2** in the first round and **this role will stay the same in all rounds**.

The **first player to click on "Take"** ends that round and **determines the payoffs of both players**, which are the payoffs **shown at the node where he/she clicks on "Take".**

**You will be paid the sum of the payoffs from all rounds.**

You will play **10 rounds.**

In **each round** you will be **matched with a different person**.

Do you have any questions at this time? If so, please raise your hand to notify the experimenter. ONCE YOU CLICK "Finish Reading" YOU WILL EXIT THE INSTRUCTIONS, SO PLEASE ASK ANY QUESTIONS YOU MAY HAVE FIRST. When you have finished reading, and have asked any questions you might have, please click "Finish Reading".

## *INSTRUCTIONS (T3)*

This is an experiment in the economics of decision-making.

During the experiment you will be participating in a decision procedure in which the participants **take turns** in **choosing between two actions**. Those actions are **"Take" or "Pass"**.

If the participant whose turn it is to choose were to select the action **"Take"**, that would **end the decision procedure**. On the other hand, if the participant whose turn it is to choose were to decide to **"Pass"**, then the **proceedings** would **continue**.

**In order to choose Take**, you would need to click on the arrow labeled Take in the tree diagram **before 10 seconds have elapsed.**

**In order to choose Pass**, you would wait for the computer to automatically enter a choice of Pass; the computer will do this after 10 seconds have elapsed.

You will be **randomly selected to be either Player 1 or Player 2**. You will **keep your role** as Player 1 or Player 2 **throughout the experiment**.

Player 1 will always be the first person to choose Pass or Take at the first node. If Player 1 chooses to Pass, Player 2 will get the next turn to Take or Pass.

*LOOKING AT DECISION OPPORTUNITIES FOR PLAYER 1 AT NODE 1*

If you are randomly selected to be Player 1 then the below tree diagram shows your available actions and payoffs. The label (1 or 2) at each node indicates who will make a decision at that node.

Player 1's payoff at a given branch of the tree is the top number; it is outlined by a rectangle in the below tree diagram. The other number, not in the rectangle, is the other person's payoff.

In order for **Player 1** to **Take at** the **first decision node**, he/she needs to click on the Take arrow; in that case **Player 1 is paid $0.40 and Player 2 is paid $0.**

In order for **Player 1** to **Pass at** the **first decision node**, he/she needs to wait for the computer to automatically enter a choice of Pass after 10 seconds have elapsed; in that case there are **no money payoffs so far and Player 2 gets the next turn** to Take or Pass.

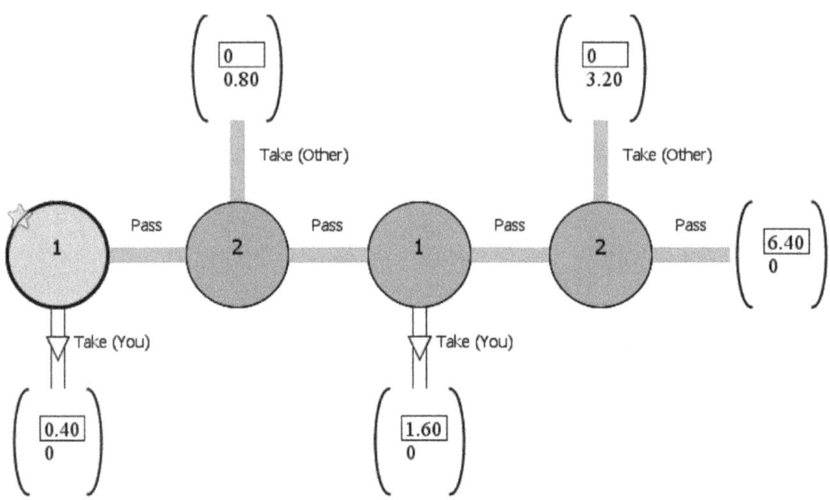

*LOOKING AT DECISION OPPORTUNITIES FOR PLAYER 2 AT NODE 2*

If you are randomly selected to be Player 2 then the below tree diagram shows your available actions and payoffs. The label (1 or 2) at each node indicates who will make a decision at that node.

Player 2's payoff at a given branch of the tree is the bottom number; it is in a rectangle in the below tree diagram. The other number, not in the rectangle, is the other person's payoff.

In order for **Player 2** to **Take at** the **second decision node**, he/she needs to click on the Take arrow; in that case **Player 2 is paid $0.80 and Player 1 is paid $0.**

In order for **Player 2** to **Pass at** the **second decision node**, he/she needs to wait for the computer to automatically enter a choice of Pass after 10 seconds have elapsed; in that case there are **no money payoffs so far and Player 1 gets the next turn** to Take or Pass.

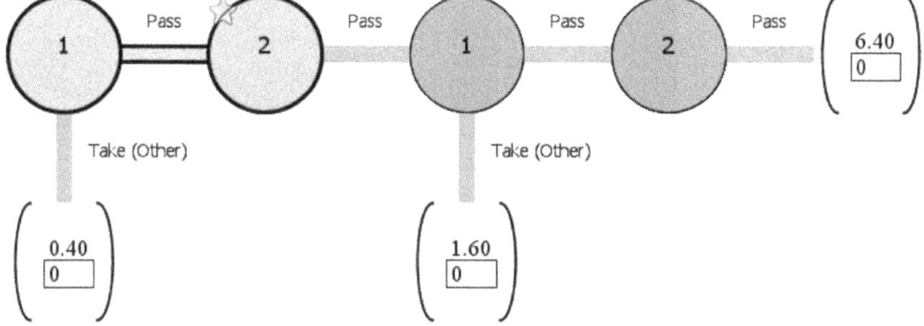

*LOOKING AT DECISION OPPORTUNITIES FOR PLAYER 1 AT NODE 3*

If **Player 1** clicks on **Take at the third decision node** then **Player 1 is paid $1.60 and Player 2 is paid $0.**

If **Player 1** chooses to **Pass at the third decision node then there are no money payoffs so far and Player 2 gets the next turn** to Take or Pass.

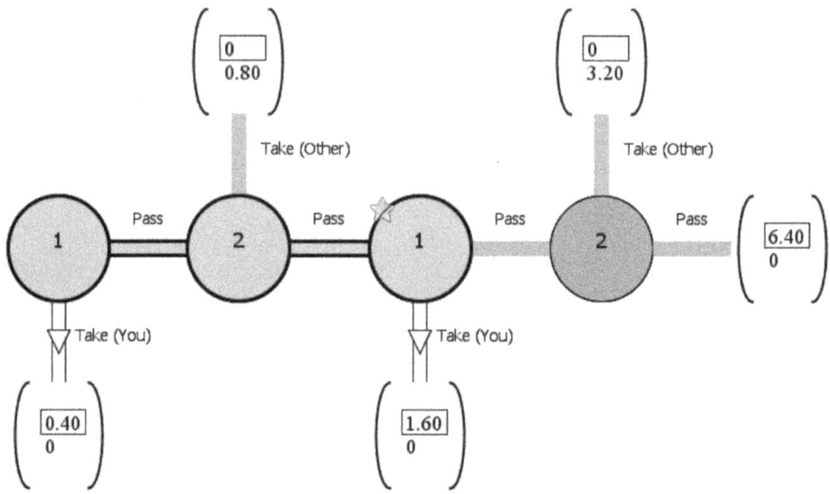

*LOOKING AT DECISION OPPORTUNITIES FOR PLAYER 2 AT NODE 4*

If **Player 2** clicks on **Take at the fourth decision node** then **Player 2 is paid $3.20 and Player 1 is paid $0.**

If **Player 2** chooses to **Pass at** the **fourth decision node** then the decision procedure ends with the payoffs at the end of the tree: **Player 2 gets paid $0 and Player 1 gets paid $6.40**.

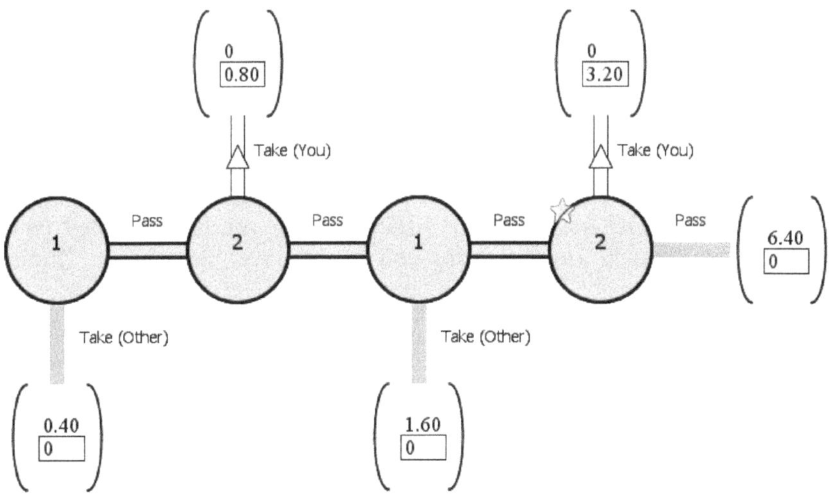

*MANY ROUNDS BUT DIFFERENT PLAYERS*

You will **participate** in the decision procedure **many times**; each participation repetition is called a "round".

You will be **randomly selected as either Player 1 or Player 2** in the first round and **this role will stay the same in all rounds**.

The **first player to click on "Take"** ends that round and **determines the payoffs of both players**, which are the payoffs **shown at the node where he/she clicks on "Take".**
**You will be paid the sum of the payoffs from all rounds.**

You will play **10 rounds.**

In **each round** you will be **matched with a different person**.

Do you have any questions at this time? If so, please raise your hand to notify the experimenter. ONCE YOU CLICK "Finish Reading" YOU WILL EXIT THE INSTRUCTIONS, SO PLEASE ASK ANY QUESTIONS YOU MAY HAVE FIRST. When you have finished reading, and have asked any questions you might have, please click "Finish Reading".

*INSTRUCTIONS (T4)*

This is an experiment in the economics of decision-making.

You will have the opportunity to engage in 10 rounds of a decision procedure.

You will be participating in a decision procedure in which the participants take turns in deciding whether to buy a fictitious item with known value of $6.40.

If you BUY THE ITEM, your earnings will be given by:

EARNINGS = $6.40–PURCHASE PRICE.

If you DO NOT BUY THE ITEM, your EARNINGS WILL BE ZERO for that round.

**If the participant whose turn it is clicks on the PURCHASE button that would end the decision procedure and he/she would buy the item at the price showing on the price clock**.

On the other hand, **if the participant whose turn it is does NOT click on the PURCHASE button then the opportunity to purchase the item passes to the other person** after a 10 second delay.

You will be **randomly selected to be either Player 1 or Player 2**. You will **keep your role** as Player 1 or Player 2 **throughout the experiment**.

Player 1 will always be the first person to have an opportunity to purchase the item. If Player 1 chooses not to purchase the item then Player 2 will get the next turn.

If Player 2 chooses NOT to purchase the item then Player 1 will get the next turn.

The current price at which a purchase can be made is displayed on the price clock.

The price on the clock ticks down every 10 seconds until either Player 1 or Player 2 clicks on the PURCHASE button and thereby purchases the item at the price displayed on the price clock at that time or the price reaches zero (in which case Player 1 will get the item).

*LOOKING AT THE FIRST PRICE CLOCK TICK*

If you have been randomly selected to be Player 2 then you cannot purchase the item at the first price clock tick; your PURCHASE button is not active on this tick.

If you have been randomly selected to be Player 1 then the first price at which you are able to purchase the item is $6.00.

If Player 1 wants to purchase the item for $6.00 then he/she should click on the PURCHASE button.

If Player 1 does NOT want to purchase the item for $6.00 then he/she DO NOTHING and the purchase opportunity will pass to Player 2 after 10 seconds.

**Player 1**

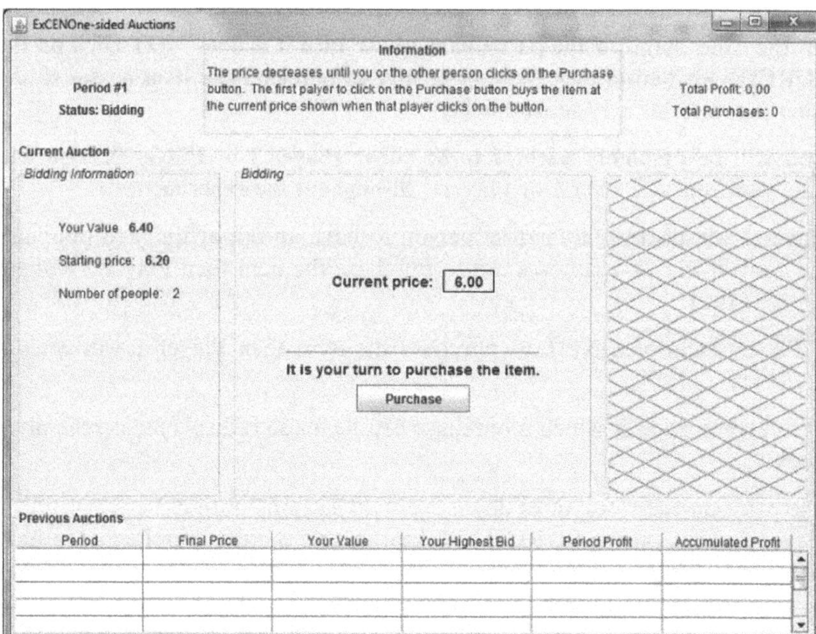

**Player 2**

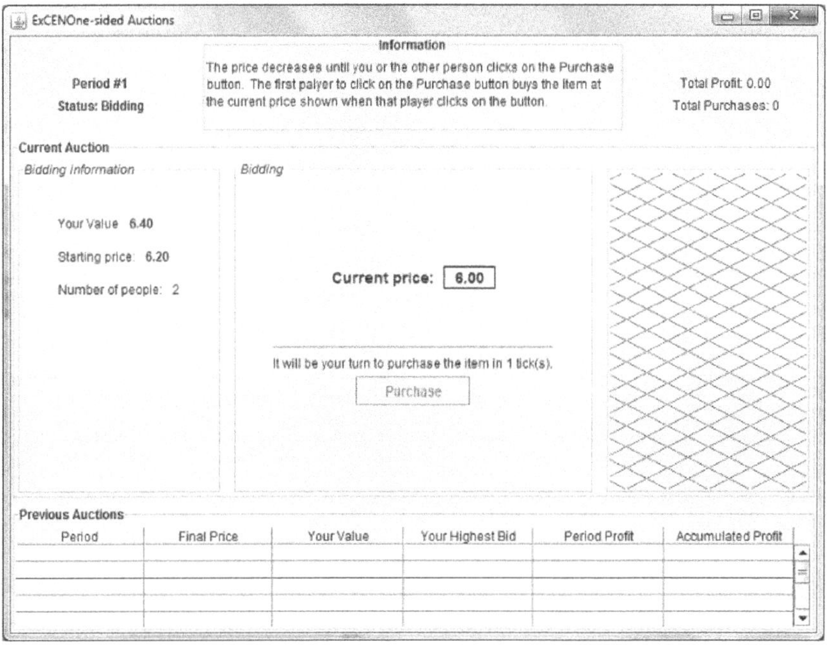


*LOOKING AT THE SECOND PRICE CLOCK TICK*

Player 1 cannot purchase the item at the second price clock tick; his/her PURCHASE button is not active on this tick.

If Player 1 did not earlier purchase the item for $6.00 then Player 2 gets an opportunity to purchase for $5.60.

If you are Player 2 and want to purchase the item for $5.60 then click on the PURCHASE button.

If Player 2 does NOT want to purchase the item for $5.60 then he/she should DO NOTHING and the purchase opportunity will pass to Player 1 after 10 seconds.

# Player 1

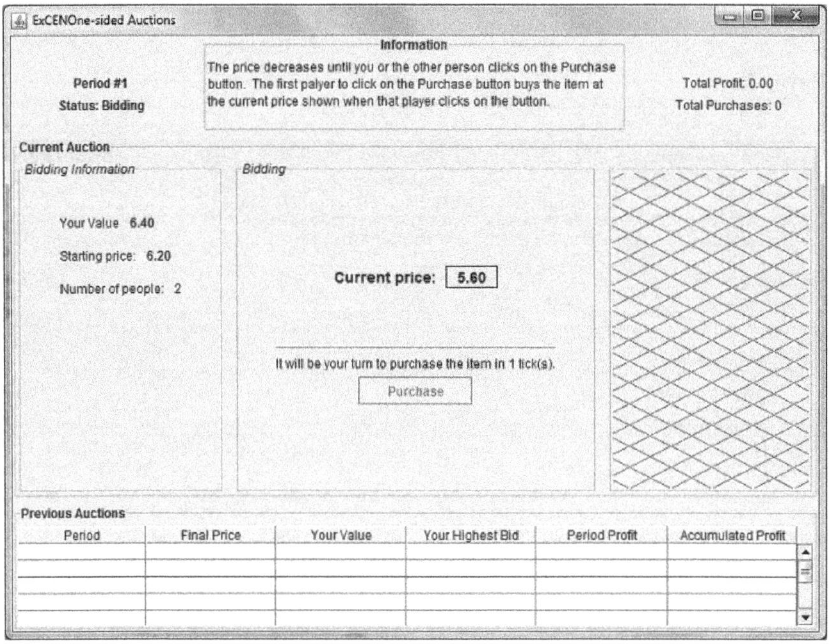


# Player 2

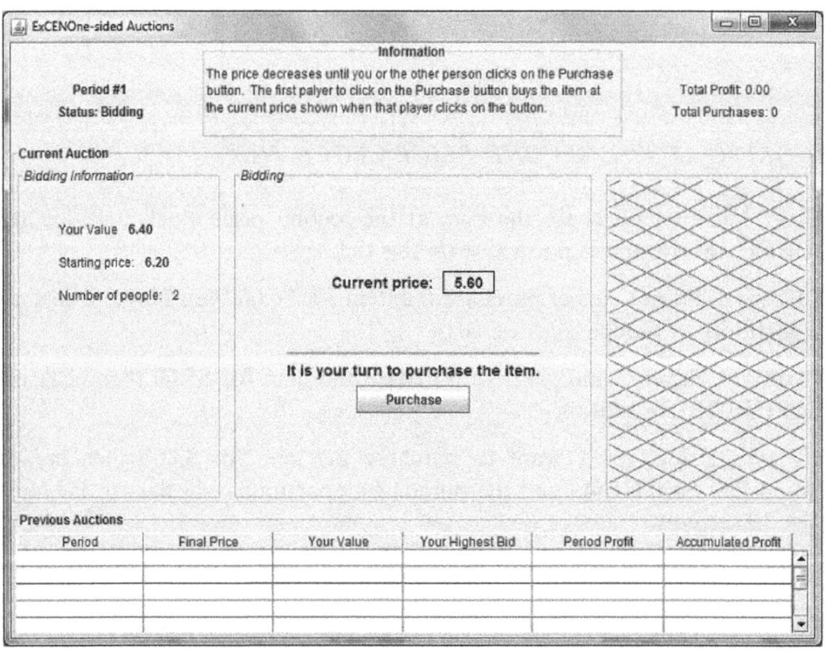

*LOOKING AT THE THIRD PRICE CLOCK TICK*

Player 2 cannot purchase the item at the third price clock tick; his/her PURCHASE button is not active on this tick.

If Player 2 did not earlier purchase the item for $5.60 then Player 1 gets an opportunity to purchase the item for $4.80.

If you are Player 1 and want to purchase the item for $4.80 then click on the PURCHASE button.

If Player 1 does NOT want to purchase the item for $4.80 then he/she should DO NOTHING and the purchase opportunity will pass to Player 2 after 10 seconds.

**Player 1**

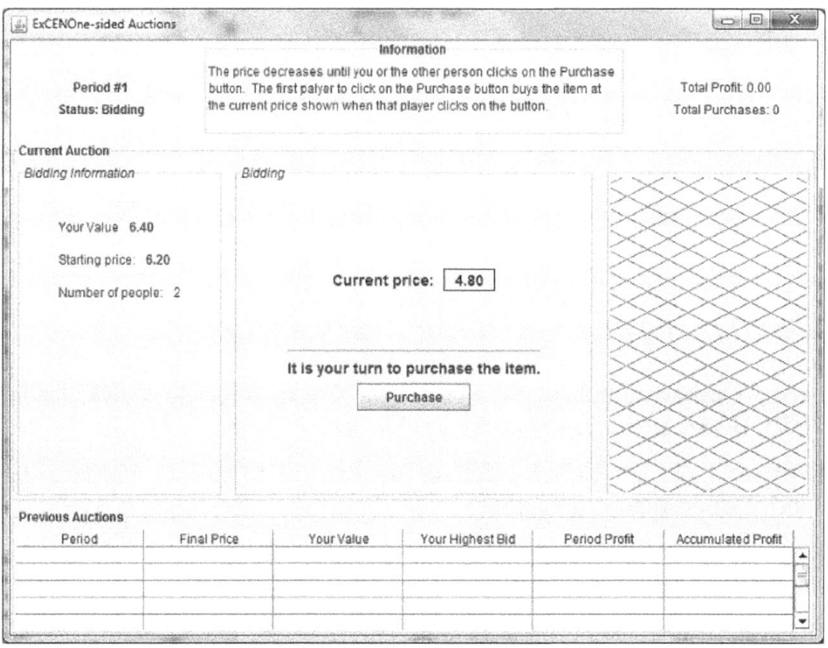

**Player 2**

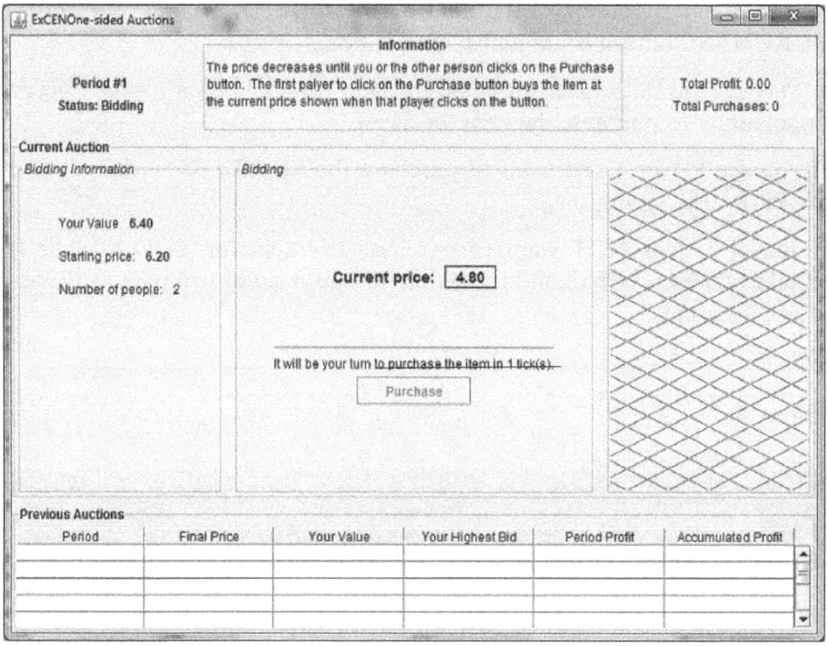


*LOOKING AT THE FOURTH PRICE CLOCK TICK*

Player 1 cannot purchase the item at the fourth price clock tick; his/her PURCHASE button is not active on this tick.

If Player 1 did not earlier purchase the item for $4.80 then Player 2 gets another opportunity to purchase.

If you are Player 2 and want to purchase the item for $3.20 then click on the PURCHASE button.

If Player 2 does NOT want to purchase the item for $3.20 then he/she should DO NOTHING and the purchase opportunity will pass to Player 1 after 10 seconds.

Player 1

Player 2

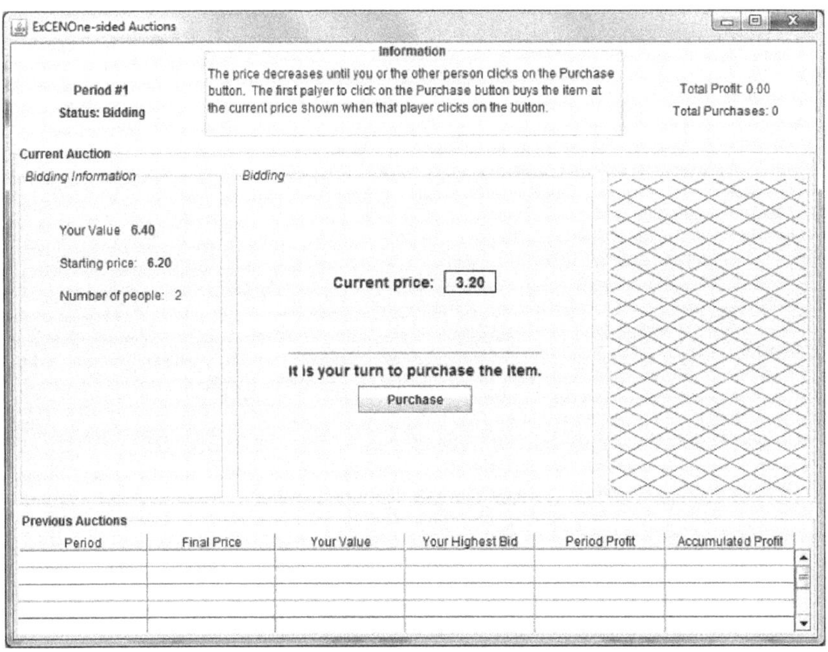

*LOOKING AT THE FIFTH PRICE CLOCK TICK*

Player 2 cannot purchase the item at the fifth price clock tick; his/her PURCHASE button is not active on this tick.

If Player 2 did not earlier purchase the item for \$3.20 then Player 1 gets to purchase the item for \$0.

If you are Player 1 and want to purchase the item for \$0 then you can click on the PURCHASE button or do nothing. **In either case, you will purchase the item for \$0.**

**Player 1**

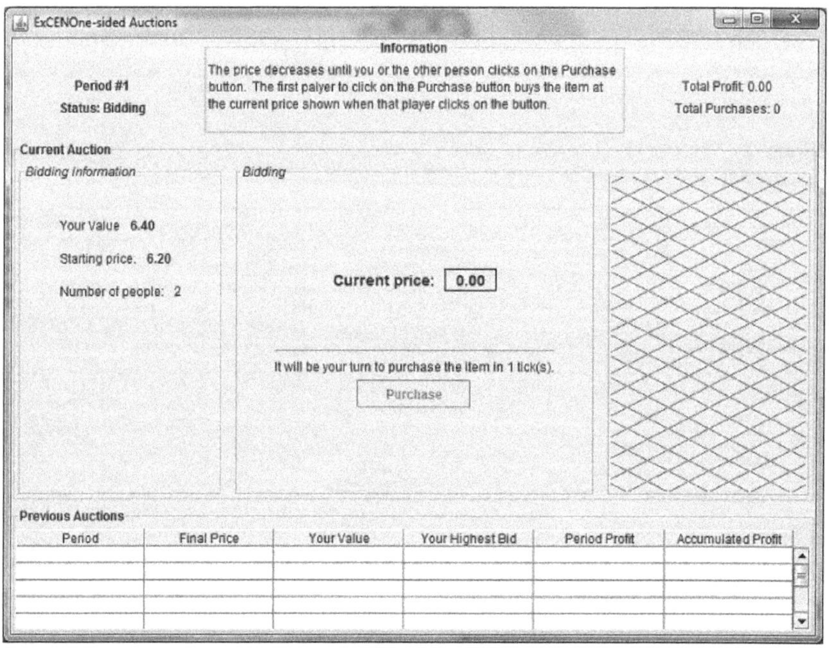

# Player 2

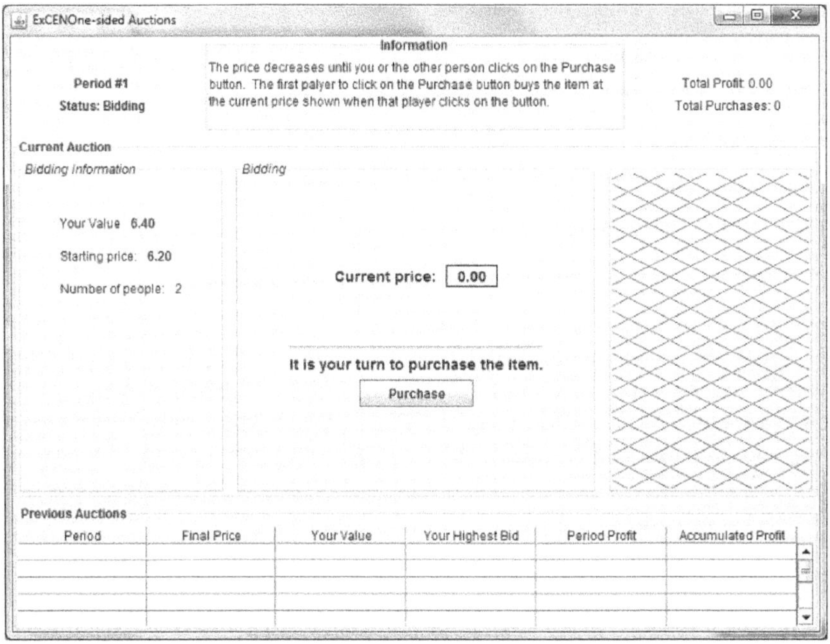
ExCENOne-sided Auctions
Information
The price decreases until you or the other person clicks on the Purchase button. The first palyer to click on the Purchase button buys the item at the current price shown when that player clicks on the button.
Period #1
Status: Bidding
Total Profit 0.00
Total Purchases: 0
Current Auction
Bidding Information
Your Value 6.40
Starting price: 6.20
Number of people: 2
Bidding
Current price: 0.00
It is your turn to purchase the item.
Purchase
Previous Auctions
Period
Final Price
Your Value
Your Highest Bid
Period Profit
Accumulated Profit

*MANY ROUNDS BUT DIFFERENT PLAYERS*

You will **participate** in the decision procedure **many times**; each participation repetition is called a "round".

You will be **randomly selected as either Player 1 or Player 2** in the first round and **this role will stay the same in all rounds**.

The **first player to click on PURCHASE** ends that round.

**You will be paid the sum of your payoffs from all rounds.**

You will play **10 rounds.**

In **each round** you will be **matched with a different person**.

Do you have any questions at this time? If so, please raise your hand to notify the experimenter. ONCE YOU CLICK "Finish Reading" YOU WILL EXIT THE INSTRUCTIONS, SO PLEASE ASK ANY QUESTIONS YOU MAY HAVE FIRST. When you have finished reading, and have asked any questions you might have, please click "Done

# THE IMPACT OF FINANCIAL HISTORIES ON INDIVIDUALS AND SOCIETIES: A REPLICATION OF AND EXTENSION OF BERG ET AL. (1995)

Xu Jiang, Radhika Lunawat and Brian Shapiro

## ABSTRACT

*We replicate and extend the social history treatment of the Berg, Dickhaut, and McCabe (1995) investment game, to further document how the reporting of financial history influences how laboratory societies organize themselves over time. We replicate Berg et al. (1995) by conducting a No History and a Financial History session to determine whether a report summarizing the financial transactions of a previous experimental session will significantly reduce entropy in the amounts sent by Investors and returned by Stewards in the investment game, as Berg et al. (1995) found. We extend Berg et al. (1995) in two ways. First, we conduct a total of five sessions (one No History and four Financial History sessions). Second, we introduce Shannon's (1948) measure of entropy from information theory to assess whether the introduction of financial transaction history reduces the amount of dispersion in the*

**Replication in Experimental Economics**
**Research in Experimental Economics, Volume 18, 95–135**

**ISSN: 0193-2306/doi:10.1108/S0193-230620150000018004**

*amounts invested and returned across generations of players. Results across sessions indicate that entropy declined in both the amounts sent by Investors and the percentage returned by Stewards, but these patterns are weaker and mixed compared to those in the Berg et al. (1995) study. Additional research is needed to test how initial conditions, path dependencies, actors' strategic reasoning about others' behavior, multiple sessions, and communication may mediate the impact of financial history. The study's multiple successive Financial History sessions and entropy measure are new to the investment game literature.*

**Keywords:** Trust; investment; return; financial history; societal organization; entropy

**JEL classifications:** C92; C65; D80; D02

Despite the proliferation of measurement and reporting practices, little is known about their impact on society at large. The purpose of this paper is to replicate and extend the social history treatment of the Berg, Dickhaut, and McCabe (1995) investment game, to further document how a prevalent public disclosure practice, the reporting of financial history, influences how laboratory societies organize themselves over time. While the investment game has been widely studied and replicated, the social history treatment has not been as extensively examined (for an exception, see Ortmann, Fitzgerald, & Boeing, 2000). We develop the idea that individuals and organizations may use information not only to draw inferences about their own and others' behavior, but also to restructure their environment. More specifically, we hypothesize that successive generations of laboratory societies will increasingly organize themselves over time when they receive reports of financial transactions undertaken by the preceding generation.

Theories of how communication influences the way societies organize themselves (e.g., Giddens, 1984; Habermas, 1979; Leydesdorff, 2001; Luhmann, 1986) provide a basis for our hypothesis. In these theories, societal organization is an emergent property of the norm-governed behavior of individual actors and can be mediated through various coordination and communications systems. For example, information systems that aggregate and publicly report the actions of individuals from a preceding generation can help organize the actions of individuals in a successive generation if individuals interpret the information as an indicator of the expected behavioral norms in their society (cf. Chaudhuri, Schotter, & Sopher, 2009). In this manner, the public information that individuals use to produce and reproduce their social systems are the product of social interaction among individuals in preceding generations (cf. Berg et al., 1995, pp. 132–133; Giddens, 1984; Latour, 1993).

We operationally define an increase in societal organization as a reduction in the entropy (disorder or variance) in the amounts invested and returned by Investors and Stewards across successive generations of players in an Investment Game setting. In an Investment Game, an Investor receives \$10 from the experimenter and then decides how much of the \$10 (\$0–\$10) to send to the Steward. The amount sent to the Steward then triples (\$0–\$30). Finally, the Steward decides how much of the tripled sum to send back to the Investor. In our experiment, each societal generation consists of a new set of Investor/Steward pairs. Following Berg et al. (1995), we introduce financial history by providing each successive generation with a summary of the amounts invested and returned by players in the preceding generation. We replicate Berg et al. (1995) by conducting a No History and a Financial History session in order to determine whether a report summarizing the financial transactions of a previous experimental session will significantly reduce entropy in the amounts sent by Investors and returned by Stewards, as the Berg et al. (1995) study shows. We also extend Berg et al. (1995) in two principal ways. First, in order to test our hypothesis that successive generations of laboratory societies will increasingly organize themselves when they receive a financial report of transactions from the immediately preceding generation, we conduct a total of five sessions (one No History and four Financial History sessions). Second, we introduce Shannon's (1948) measure of entropy from information theory to assess whether the introduction of financial transaction history reduces entropy (the amount of dispersion) in the distributions of amounts invested and returned across generations of players.

The relatively simple and widely used setting of the Investment Game has several attractive features for investigating the evolution of societal organization. The game is played only once, Investors and Stewards are anonymously paired, and anonymity ensures that the experimenter cannot match the identities of specific individuals to the dollar amounts sent and returned. These procedures control for alternative explanations of behavior including repeated game reputation effects, contractual precommitments, punishment threats, and experimenter demand effects (e.g., Ortmann et al., 2000). In addition, in the Investment Game and other similar games (e.g., the gift exchange game, the peasant-dictator game, and the moonlighting game), many players invest and return nonzero amounts, contrary to the assumption of pure monetary self-interest (Berg et al., 1995; Camerer & Fehr, 2002; Cox, 2004; Dickhaut & McCabe, 1997; for a meta-analysis of the Investment Game, see Johnson & Mislin, 2011). Finally, in the Investment Game setting, substantial variation exists across the actions of different players (e.g., Berg et al., 1995; Johnson & Mislin, 2011; Ortmann et al., 2000). Following Berg et al. (1995), we hypothesize that the introduction of a financial transaction history will reduce this variation by mediating what individuals expect of others and what others expect of them. In addition, we extend Berg et al. (1995) by further hypothesizing that with financial histories given to successive generations of laboratory societies, the actions of individuals will on average converge across those societies toward emerging norms of expected behavior.

In sum, our study attempts to replicate Berg et al.'s (1995) results and extends the prior investment game literature in two principal ways. First, it applies our entropy metric to the Berg et al. (1995) data. Second, it ascertains whether the introduction of a financial history has an incremental effect across multiple generations of the investment game.

The remainder of the paper is organized as follows. The first section describes the Investment Game, develops our measure of entropy, and states our hypotheses. The subsequent two sections present the experimental method and results, respectively. The concluding section discusses the implications of our results and identifies opportunities for future research.

## THEORY AND HYPOTHESES

Below we describe the Investment Game and develop our hypotheses about entropy reduction and societal organization.

### *Investment Game Structure*

The investment game is played as follows. In stage one, the subjects in Rooms A and B are each given \$10 as a show-up fee. Subjects in Room A (Investors) are given an additional \$10 and must decide how much of their \$10 to send to an anonymous counterpart in Room B. We denote this amount by $M_a$. The amount sent is then tripled to $3M_a$. In stage two, a counterpart in Room B (the Steward) is given the tripled money. The counterpart Steward must decide how much money to return, which is denoted $k_b(3M_a)$. The Investor chooses the strategy $M_a \in \{0, 1, 2, \ldots, 10\}$, while the Steward chooses the strategy:

$$k_b : \{0, 3, \ldots, 30) \rightarrow \{0, 1, \ldots, 30\}$$

which satisfies $0 \leq k_b(3M_a) \leq 3M_a$.

These strategies result in the payoffs:

$$P_a(M_a, k_b) = \$10 + (\$10 - M_a) + k_b(3M_a)$$

and

$$P_b(M_a, k_b) = \$10 + 3M_a - k_b(3M_a)$$

A subject's initial wealth is denoted $W_i$. If subjects have strictly increasing indirect utility function for wealth, given by $V_i(W_i + P_i(M_a, k_b))$ for $i$ = a, b, and each subject, $i$, maximizes $V_i(\cdot)$, then Stewards have a dominant strategy to keep all the money, that is, $k_b(3M_a) = 0$ for all $M_a$. If Investors infer their counterpart's dominant strategy, then they should send nothing, that is, $M_a = 0$. If these results obtain, societal entropy will be zero, even in the absence of information about the actions of individuals in a preceding laboratory society.

## *Hypotheses*

One hypothesis for the investment game is that subjects will make decisions consistent with the subgame perfect prediction,

$N_0 : M_a = 0$ for all a

If for some reason Investors send a positive amount (i.e., $M_a > 0$), then the dominant strategy for Stewards is as follows:

$$N_1 : \text{If } M_a > 0, \text{then } k_b(3M_a) = 0 \text{ for all b}$$

Past research in the Investment Game has found, however, that some investors trust stewards and that some stewards are trustworthy (Berg et al.,

1995; Cox, 2004; Johnson & Mislin, 2011; Ortmann et al., 2000). Consistent with these prior results, our hypotheses for the No History condition (first generation of players) in our Investment Game are as follows:

**H1.** Investors will send nonzero amounts to Stewards ($M_a > 0$ for some "a").

**H2.** Stewards who receive nonzero amounts will return nonzero amounts to Investors ($k_b(3M_a) > 0$ for some "b").

*Impact of Financial History*

The question next arises as to how much variation exists across individual behavior, and whether a simple information system that merely aggregates and reports the actions of individuals from a previous generation will reduce that variation in a successive generation. Berg et al. (1995) previously reported less variation between a No History and one subsequent Social History session. We attempt to replicate this result as well as test our general hypothesis that when financial histories are provided to successive generations of laboratory societies, the actions of individuals will on average converge across those societies toward emerging norms of expected behavior.

In our experimental setting, all generations except the first receive a financial history of transactions made by the immediately preceding generation. Sociological research suggests that communication across generations can significantly influence behavior in subsequent generations (e.g., Giddens, 1984; Habermas, 1979; Leydesdorff, 2001; Luhmann, 1986; see also Chaudhuri et al., 2009), which documented the conditions under which intergenerational communication improves coordination in the Minimum Effort Game setting. In addition, Lunawat (2013a, 2013b) used the Berg et al. (1995) setting to show how communication of private financial information can enhance trust between a manager and an investor. In our setting, a financial history summarizing the transactions of a preceding generation may provide a basis for individuals to assess the dispositions, intentions, and trustworthiness of players in a subsequent Investment Game (cf. Kramer, 1999, p. 575). For example, Investors might use the past transaction data to identify investments that are likely to yield a positive return,[1] and Stewards might infer the reasonableness of the amount invested by their Investor partners as well as the amount they are expected to return to the Investors. Alternatively, Investors and Stewards might more simply select amounts based on their conformity with perceived central tendencies in the past data, regardless of whether the central tendencies are socially beneficial.

*Using Entropy to Measure the Impact of Financial History*

To assess the impact of financial history on players' behavior, we measure the change in entropy of the amounts invested and returned by Investors and Stewards. The term entropy first arose in statistical mechanics to reflect the status of a particle system. Decreases in entropy reflect less movement in the particle and less uncertainty about its location. In a social setting, the amount of entropy reflects the amount of uncertainty about individuals' behavior. An increase in entropy means that the behavior of individuals is more uncertain, while a decrease in entropy means that the behavior is less uncertain or more orderly.

The entropy concept has been widely used in the engineering and information theory literatures (e.g., see Cover & Thomas, 1991). Sims (1998, 2003, 2005) subsequently introduced Shannon's (1948) entropy concept in the economics literature and defined the informativeness of information signals as the change in entropy between prior and posterior distributions. The entropy concept has been used to model price stickiness (Sims, 1998), flexible information acquisition (Yang, 2015), and general equilibrium (Mackowiak & Wiederholt, 2009). Following these prior studies, we use the entropy concept to measure the amount of uncertainty about individuals' behavior.

More formally, we define the amount of order in our laboratory societies as follows. Let $p_i$ denote the probability of the $i$th element being drawn from a distribution of $n$ discrete outcomes. Entropy is defined by $-\sum_i p_i \log_2 p_i$. As an example, consider a binary outcome such as the flip of a coin. A fair coin has entropy equal to 1, and a biased coin with a 0.75 probability of heads has entropy equal to 0.81.[2] The entropy metric indicates that the biased coin has more order than the fair coin.[3]

*Entropy Metric Applied to the Berg et al. (1995) Data*

To illustrate how the entropy metric can be applied and to establish continuity with the prior literature, we now perform an entropy analysis of the Berg et al. (1995) data. The Berg et al. (1995) study had two treatments, "no history" and "social history." The social history treatment included a report summarizing the decisions in the no history treatment. Below, we re-label their "social history" treatment as "financial history" to conform to the labels we use in our experiments. We compare the entropy metrics across the Berg et al. (1995) no history and financial history treatments, separately for the investor, steward, and joint data sets.

For the Investors, $\text{Entropy}_{\text{No History}} = 3.141$ and $\text{Entropy}_{\text{Financial History}} = 2.876$. We used a bootstrapping technique to determine whether the observed Entropy reduction of 0.265 is statistically significant. The null hypothesis for this test is that the amounts invested were chosen randomly. For this and all subsequent tests reported below, we generated 10,000 measures of differences in entropy. Based on the bootstrapped distribution, the observed reduction in entropy is significant ($p < .02$, one-tailed).

Since the maximum amounts returned by stewards were constrained by the tripled amounts sent by investors, we converted the amounts returned into percentages of the maximum amount that could be returned. Under this measure, more trustworthy stewards returned a higher percentage. More specifically, we analyzed the entropy in the distributions of the amounts returned by Stewards, in deciles of percentage of amounts received from Investors. For these Steward distributions, $\text{Entropy}_{\text{No History}} = 3.086$ and $\text{Entropy}_{\text{Financial History}} = 2.709$. A bootstrapped distribution indicates that the observed Entropy difference of 0.377 is statistically significant ($p < .0001$, one-tailed).

Finally, the Berg et al. (1995) no history and financial history data as a whole tended to cluster on outcome pairs. We examine the joint densities of Investor and Steward behavior to test whether the observed clustering is statistically significant. This test must take into account that the No History condition has a smaller sample size than the Financial History condition. Ceteris paribus, smaller sample sizes yield smaller maximum levels of entropy.[4] The entropy formula for the No History condition is:

$$-\sum_i \sum_j \frac{f_{i,j}}{32} \log_2 \frac{f_{i,j}}{32}$$

where $f_{i,j}$ denotes the frequency in category "*i*, *j*." For example, "3,2" denotes 3 dollars invested and 2 returned by the steward. A similar formula applies to the Financial History condition, with 28 possible observations. The calculated entropies are 4.688 and 4.227 for the No History and Financial History conditions, respectively, for an overall decrease in entropy of 0.461 across the two conditions. Based on a bootstrapped distribution of the difference in entropies across the two conditions, the 0.461 reduction in entropy is marginally significant at the .055 level (one-tailed).

In sum, the Berg et al. (1995) data indicate that the introduction of financial history significantly reduced entropy in the distributions of amounts sent and amounts returned, and marginally reduced entropy in the joint distributions. Below we attempt to replicate this result, as well as test whether financial histories will have an incremental effect over subsequent generations of laboratory societies. To address the latter, we gathered new data over five societal generations to test the predictive power of our hypothesis that financial histories provide a basis for developing order in societies. Participants in Session I (the first generation) received no financial history, whereas participants in Sessions II–V (the four subsequent generations) received a report summarizing the financial history of the immediately preceding session. Thus, Sessions I and II replicate the original Berg et al. (1995) study, and Sessions III–V extend that study.

Based on our preceding discussion of social theory and our entropy analyses of the Berg et al. (1995) data, the following hypotheses predict that the distributions of the amounts invested, returned, and their joint distributions will exhibit more order (less entropy) across generations of players:

**H3.** Financial transaction histories will produce more order (less entropy) in the Investor distributions across generations.

**H4.** Financial transaction histories will produce more order (less entropy) in the Steward distributions across generations.

**H5.** Financial transaction histories will produce more order (less entropy) in the joint Investor/Steward distributions across generations.

## METHOD

### *Subjects*

Subjects were recruited and participated in the experimental sessions in two groups separated by time and geography. Group 1 subjects were recruited in October 2007 from the subject pool in the CIRANO Research Center, Montreal, Canada. The CIRANO subject pool was comprised mostly of undergraduate and graduate students from the universities that participated in the research center. Due to a nondiscrimination clause governing subject recruiting, a few Group 1 subjects were nonstudent working professionals from the downtown Montreal area. Group 2 subjects were mostly undergraduate students recruited in February and March 2015 from the Experimental Social Sciences Lab at the University of California – Irvine.

The same experimental procedures were administered to both subject groups. Subjects received $10 for showing up. Subjects also could earn up to an additional $30, depending on their decisions and the decisions of their anonymous partner. Subjects were told to report directly to either Room A or Room B. Room A subjects served as Investors, and Room B subjects served as Stewards. A monitor was randomly chosen in each room after 10 subjects arrived. Monitors received $5, in addition to their $10 show-up fee.

Five sessions consisting of nine pairs of subjects and two monitors were run for both Group 1 and Group 2. The first session was the "No History" condition, in which subjects did not receive any financial history from a previous experimental session. The other four sessions received a financial history from the immediately preceding experimental session. The Appendix provides the instructions for both investors and stewards, for both the non-history and history conditions. Table 1 (Table 2) summarizes the results for each of the nine pairs of Group 1 (Group 2) subjects in each session.

## *Materials*

Each subject was given a consent form and set of instructions. The consent form emphasized that (a) participation in the experiment will not affect participants' relationship with the University or affect their course grades, (b) participation is voluntary, (c) all data will remain confidential, and (d) compensation consists of a payment for showing up plus or minus amounts sent and received to anonymous counterparts in the experiment. The instructions were read aloud by the experimenter.

## *Procedures*

The Appendix provides the instructions, including the procedures taken to assure anonymity and confidentiality. Each subject performed the following steps:

A. Subjects were randomly assigned to report directly to either Room A or Room B. One subject in each room was randomly selected to serve as a monitor (called Monitor A or Monitor B, for Room A and Room B, respectively). Monitors verified that the experimenters followed the experimental instructions.

B. Subjects read the consent form, were asked questions about the form, were given an opportunity to ask questions about the form, and then signed the form if they wished to participate in the experiment.
C. An experimenter in Room A read the experimental instructions out loud in Room A, and another experimenter in Room B read the instructions out loud in Room B. Subjects in both rooms were given hard copies of the instructions and were asked to read along silently. Toward the end of the instructions, a summary financial history was given to all but the first generation of players.
D. One at a time, Room A subjects (Investors) were called to a private area in Room A, were randomly given an envelope containing $10, opened the envelope, and decided whether to send all, some, or none of $10 to their Room B counterparts (Stewards). Room A subjects put the amount of money they wish to send in an envelope and pocketed the rest. Room A subjects also pocketed a specially marked key to be used later.
E. After all Room A subjects performed step D, the Room A monitor took the envelopes to a recorder who was located in a hallway. The recorder recorded the amount sent and tripled the investment. The Room B monitor was then called to bring the envelopes to Room B, and Monitor A returned to Room A.
F. One at a time, Room B subjects were called to a private area in Room B, were randomly given an envelope, opened the envelope, decided how much money to send back to their counterpart in Room A, left that amount of money in the envelope, and pocketed the rest. Afterward, each Room B subject was paid a $10 show-up fee and was asked to leave the building.
G. The Room B monitor took the envelopes to the recorder in the hallway. At this time the amounts sent back were recorded. The Room A monitor was called to put the envelopes into mailboxes marked with letters corresponding to letters on the envelopes.
H. One at a time, Room A subjects went to Room C and opened the mailbox with a lettered key corresponding to the letter on mailbox. After opening the appropriate mailbox, Room A subjects took out the envelope, removed the money, and dropped the key in a box. Afterward, Room A subjects were paid their $10 show-up fee and were then asked to leave the building.
I. The monitors were paid $15 for their participation.

### *Comparison with Berg et al. (1995)*

Besides the difference in subject pools, our experimental setting differs from the Berg et al. (1995) experiment in one respect. In the investment games reported in Berg et al. (1995), all subjects were given a $10 show-up fee. Room A subjects then decided how much of their $10 show-up fee to send to their anonymous counterpart in Room B, and Room B subjects pocketed their show-up fee. Our experiment followed the same procedures, except that Room A subjects also received an additional $10 show-up fee.

## RESULTS

Below we report the results for each hypothesis, for Group 1 followed by Group 2. We also report results where we combine both groups in footnotes.

### *Amounts Invested and Returned*

Participants in Session I have no history, while participants in Sessions II–V have financial history because they have access to the results of prior sessions. H1 predicted that investors would send nonzero amounts to stewards. For Group 1, the average amount invested ranged from $4.67 in Session I to $7.00 in Session V (Table 1), with an overall average of $5.45. One-sided Wilcoxon rank-sum tests indicate that in all five sessions, the amounts invested differed significantly from zero with a one-sided *p*-value <.023. For Group 2, the average amount invested ranged from $5.22 in Session I to $4.11 in Session V (Table 2), with an overall average of $2.69. One-sided Wilcoxon rank-sum tests indicate that in all five sessions, the amounts invested differed significantly from zero with a one-sided *p*-value <.054. These results support H1. As a direct comparison with Berg et al. (1995), we also compare the amounts sent and returned between Session I (i.e., the generation without financial history) and Session II (i.e., the immediate generation that observes financial history). For Group 1, the average amounts sent for Sessions I and II are $4.67 and $5.22, respectively. For Group 2, the average amounts sent for Sessions I and II are $5.00 and $2.67, respectively. One-sided Wilcoxon rank-sum tests indicate that there is no statistical difference between the amounts sent of the two

***Table 1.*** Group 1 Amounts Sent and Returned, by Experimental Session.

| Session I | | | Session II | | |
|---|---|---|---|---|---|
| Subject pair | Sent | Returned | Subject pair | Sent | Returned |
| A | 6 | 0 | A | 6 | 9 |
| B | 5 | 10 | B | 2 | 1 |
| C | 0 | 0 | C | 0 | 0 |
| D | 5 | 0 | D | 6 | 3 |
| E | 4 | 0 | E | 10 | 12 |
| F | 4 | 4 | F | 4 | 5 |
| G | 8 | 16 | G | 2 | 1 |
| H | 0 | 0 | H | 10 | 5 |
| I | 10 | 0 | I | 5 | 15 |
| Averages: | $4.67 | $4.29 | Averages: | $5.00 | $6.38 |
| Session III | | | Session IV | | |
| A | 0 | 0 | A | 10 | 0 |
| B | 5 | 8 | B | 5 | 8 |
| C | 5 | 10 | C | 0 | 0 |
| D | 1 | 1 | D | 5 | 10 |
| E | 4 | 5 | E | 3 | 0 |
| F | 4 | 0 | F | 10 | 21 |
| G | 5 | 8 | G | 5 | 15 |
| H | 7 | 7 | H | 6 | 5 |
| I | 10 | 15 | I | 10 | 8 |
| Averages: | $4.56 | $6.75 | Averages: | $6.00 | $8.38 |
| Session V | | | | | |
| A | 5 | 0 | | | |
| B | 5 | 3 | | | |
| C | 5 | 10 | | | |
| D | 10 | 13 | | | |
| E | 10 | 15 | | | |
| F | * | * | | | |
| G | 6 | 9 | | | |
| H | 10 | 20 | | | |
| I | 5 | 5 | | | |
| Averages: | $7.00 | $9.38 | | | |

Investor averages include $0 amounts invested. Steward averages exclude observations where $0 was received.
* Missing data.

sessions for both groups, although the amount invested appears to be increasing in Group 1 but decreasing in Group 2. With respect to the amount returned, the average amounts returned for Sessions I and II are \$4.29 and \$5.25, respectively. For Group 2, the average amounts returned are \$6.28 and \$2.71, respectively. Again, one-sided Wilcoxon rank-sum tests indicate that there is no statistical difference between the amounts returned for the two sessions for both groups, although the amount returned appears to be increasing in Group 1 but decreasing in Group 2.

H2 predicted that stewards who receive nonzero amounts will return nonzero amounts. Therefore, in order to test H2 we removed observations where the stewards received \$0. In Group 1, the average amounts returned by stewards for those who received nonzero amounts increased monotonically from \$4.29 in Session I (no history) to \$9.38 in Session V (financial history), with an overall average of \$7.04 (Table 1). Excluding cases where stewards received \$0, one-sided Wilcoxon rank-sum tests for amounts that are converted into percentages of the maximum amount that could be returned yielded one-sided $p$-values <.023 for all sessions other than Session I. In Session I, the Wilcoxon rank-sum test yielded a one-sided $p$-value of .174. The latter results may be attributed to the relatively large number of Stewards who received positive amounts but returned \$0 in Session I. In Group 2, however, the average amounts returned by stewards decreased initially from \$5.25 in Session I (no history) to \$1.29 in Session III before rising to \$3.76 in Session V, for an overall average of \$3.50 (Table 2). Excluding cases where stewards received \$0, one-sided Wilcoxon rank-sum tests for amounts that are converted into percentages of the maximum amount that could be returned yielded one-sided $p$-values <.10 for Sessions I, IV, and V. For Sessions II and III, however, the one-sided $p$-value is .10 and .37, respectively. Overall, the results from Group 1 and Group 2 support H2, but the Group 2 results are weaker. Again, as a direct comparison with Berg et al. (1995), we also compare the percentages of the maximum amount that could be returned between Session I (i.e., the generation without financial history) and Session II (i.e., the immediate generation that observes financial history). For Group 1, the average percentage returned for Sessions I and II are 0.238 and 0.366, respectively. For Group 2, the average percentage returned for Sessions I and II are 0.275 and 0.371, respectively. One-sided Wilcoxon rank-sum tests for both groups indicate that there is no statistical difference between the percentage of the amounts sent across the two sessions, although the percentage invested appears to be increasing in both groups.

***Table 2.*** Group 2 Amounts Sent and Returned, by Experimental Session.

| Session I | | | Session II | | |
|---|---|---|---|---|---|
| Subject pair | Sent | Returned | Subject pair | Sent | Returned |
| A | 1 | 1 | A | 8 | 10 |
| B | 5 | 8 | B | 0 | 0 |
| C | 7 | 1 | C | 2 | 0 |
| D | 8 | 0 | D | 1 | 1 |
| E | 4 | 3 | E | 5 | 0 |
| F | 10 | 18 | F | 0 | 0 |
| G | 5 | 10 | G | 5 | 7 |
| H | 0 | 0 | H | 1 | 0 |
| I | 7 | 1 | I | 2 | 1 |
| Averages: | $5.22 | $5.25 | Averages: | $2.67 | $2.71 |
| Session III | | | Session IV | | |
| A | 6 | 0 | A | 0 | 0 |
| B | 3 | 4 | B | 9 | 10 |
| C | 1 | 0 | C | 3 | 3 |
| D | 2 | 0 | D | 0 | 0 |
| E | 7 | 5 | E | 0 | 0 |
| F | 0 | 0 | F | 3 | 5 |
| G | 6 | 0 | G | 5 | 5 |
| H | 0 | 0 | H | 0 | 0 |
| I | 2 | 0 | I | 3 | 0 |
| Averages: | $3.00 | $1.29 | Averages: | $2.56 | $4.60 |
| Session V | | | | | |
| A | 1 | 1 | | | |
| B | 10 | 10 | | | |
| C | 0 | 0 | | | |
| D | 6 | 0 | | | |
| E | 6 | 0 | | | |
| F | 8 | 8 | | | |
| G | 0 | 0 | | | |
| H | 3 | 3 | | | |
| I | 3 | 4 | | | |
| Averages: | $4.11 | $3.71 | | | |

Investor averages include $0 amounts invested. Steward averages exclude observations where $0 was received.

Although we did not predict the effect that financial history might have on the amounts sent and returned, in Group 1 we did observe a successive increase in the amounts sent by investors in three of the four financial history conditions, and a successive increase in the amounts returned by Stewards in all four financial history conditions. These Group 1 results suggest that financial history can yield socially beneficial outcomes. In Group 2, however, both the amounts sent and the amounts returned are largest in Session 1 (No Financial History); the amounts sent and returned actually decreased in Sessions II and III, and only partly rebounded in Sessions IV and V. Overall, these results provide mixed evidence that financial histories can foster socially beneficial outcomes in the Investment Game setting.

### *Influence of Financial History on Entropy*

*Investor Behavior*

H3 predicted that the introduction of financial history would reduce the entropy of amounts sent by investors. Panel A of Table 3 and Panel A of Table 4 list the number of investors, tabulated by amount invested and experimental session, for Group 1 and Group 2, respectively. Figs. 1(a) and (b) display histograms of the amounts invested in each of the experimental sessions for Group 1 and Group 2, respectively. A visual inspection of Fig. 1(a) indicates that the introduction of a financial history reduced the variance in the Group 1 amounts invested across the experimental sessions, where the amounts in Session I are more dispersed than in Session V. More formally, $\text{Entropy}_{\text{session I}} = 2.503$ and $\text{Entropy}_{\text{session V}} = 1.406$, indicating a reduction in entropy of 1.098. Fig. 2(a) displays the Group 1 bootstrapping results for the amounts invested. Based on the bootstrapped distribution, the observed 1.098 reduction in entropy for the amounts invested is statistically significant ($p < .01$, one-tailed). These Group 1 results support H3, and are consistent with our entropy analysis of the Berg et al. (1995) data. Again, as a direct comparison with Berg et al. (1995), we also compared the reduction in entropy for the amounts invested between Sessions I and II. We have $\text{Entropy}_{\text{session I}} = 2.503$ and $\text{Entropy}_{\text{session II}} = 2.503$, indicating a reduction in entropy of 0. Bootstrapping showed that this entropy reduction of 0 is statistically insignificant. Thus, while Berg et al. (1995) shows that one generation of financial history is able to reduce entropy, our results indicate that one generation of financial history may not be enough.

***Table 3.*** Number of Group 1 Investors, Listed by Experimental Session and Amounts Sent and Returned.

Panel A. Number of Group 1 investors, by experimental session and amount sent.

| Amounts sent | Session I | Session II | Session III | Session IV | Session V |
|---|---|---|---|---|---|
| 0 | 2 | 1 | 1 | 1 | 0 |
| 1 | 0 | 0 | 1 | 0 | 0 |
| 2 | 0 | 2 | 0 | 0 | 0 |
| 3 | 0 | 0 | 0 | 1 | 0 |
| 4 | 2 | 1 | 2 | 0 | 0 |
| 5 | 2 | 1 | 3 | 3 | 4 |
| 6 | 1 | 2 | 0 | 1 | 1 |
| 7 | 0 | 0 | 1 | 0 | 0 |
| 8 | 1 | 0 | 0 | 0 | 0 |
| 9 | 0 | 0 | 0 | 0 | 0 |
| 10 | 1 | 2 | 1 | 3 | 3 |
| Entropy | 2.503 | 2.503 | 2.419 | 2.113 | 1.406 |

Panel B. Number of Group 1 stewards, by experimental session and amounts returned in deciles of percentage of amounts received from investors. Observations where investors sent $0 are excluded.

| Deciles of percentage | Session I | Session II | Session III | Session IV | Session V |
|---|---|---|---|---|---|
| 0 | 4 | 0 | 1 | 2 | 1 |
| (0%, 10%] | 0 | 0 | 0 | 0 | 0 |
| (10%, 20%] | 0 | 4 | 0 | 0 | 1 |
| (20%, 30%] | 0 | 0 | 0 | 2 | 0 |
| (30%, 40%] | 1 | 1 | 2 | 0 | 1 |
| (40%, 50%] | 0 | 2 | 2 | 0 | 3 |
| (50%, 60%] | 0 | 0 | 2 | 1 | 0 |
| (60%, 70%] | 2 | 0 | 1 | 2 | 2 |
| (70%, 80%] | 0 | 0 | 0 | 0 | 0 |
| (80%, 90%] | 0 | 0 | 0 | 0 | 0 |
| (90%, 100%] | 0 | 1 | 0 | 1 | 0 |
| Entropy | 1.379 | 1.750 | 2.250 | 2.250 | 2.156 |

Similarly, a visual inspection of Fig. 1(b) indicates that the introduction of a financial history reduced the variance in the Group 2 amounts invested across the experimental sessions, where the amounts in Session I are more dispersed than in Session V. More formally, $\text{Entropy}_{\text{session I}} = 2.725$ and $\text{Entropy}_{\text{session V}} = 2.503$, indicating a reduction in entropy of 0.222. Fig. 2(b) displays the Group 2 bootstrapping results for the amounts invested. The bootstrapped distribution indicates that the observed 0.222 reduction in entropy for the Group 2 amounts invested is not statistically significant

***Table 4.*** Number of Group 2 Investors, Listed by Experimental Session and Amounts Sent and Returned.

Panel A. Number of Group 2 investors, by experimental session and amount sent.

| Amounts Sent | Session I | Session II | Session III | Session IV | Session V |
|---|---|---|---|---|---|
| 0 | 3 | 2 | 2 | 4 | 2 |
| 1 | 2 | 2 | 1 | 0 | 1 |
| 2 | 0 | 2 | 2 | 0 | 0 |
| 3 | 1 | 0 | 1 | 3 | 2 |
| 4 | 3 | 0 | 0 | 0 | 0 |
| 5 | 2 | 2 | 0 | 1 | 0 |
| 6 | 3 | 0 | 2 | 0 | 2 |
| 7 | 1 | 0 | 1 | 0 | 0 |
| 8 | 1 | 1 | 0 | 0 | 1 |
| 9 | 0 | 0 | 0 | 1 | 0 |
| 10 | 1 | 0 | 0 | 0 | 1 |
| Entropy | 2.725 | 2.281 | 2.503 | 1.753 | 2.503 |

Panel B. Number of Group 2 stewards, by experimental session and amounts returned in deciles of percentage of amounts received from investors. Observations where investors sent $0 are excluded.

| Deciles of Percentage | Session I | Session II | Session III | Session IV | Session V |
|---|---|---|---|---|---|
| 0 | 1 | 3 | 5 | 1 | 2 |
| (0%, 10%] | 2 | 0 | 0 | 0 | 0 |
| (10%, 20%] | 0 | 1 | 0 | 0 | 0 |
| (20%, 30%] | 1 | 0 | 1 | 0 | 0 |
| (30%, 40%] | 1 | 1 | 0 | 3 | 4 |
| (40%, 50%] | 0 | 2 | 1 | 0 | 1 |
| (50%, 60%] | 2 | 0 | 0 | 1 | 0 |
| (60%, 70%] | 1 | 0 | 0 | 0 | 0 |
| (70%, 80%] | 0 | 0 | 0 | 0 | 0 |
| (80%, 90%] | 0 | 0 | 0 | 0 | 0 |
| (90%, 100%] | 0 | 0 | 0 | 0 | 0 |
| Entropy | 2.5 | 1.842 | 1.149 | 1.371 | 1.379 |

at conventional significance levels. In sum, the group 1 results are consistent with H3 and our entropy analysis of the Berg et al. (1995) investor amounts, but the Group 2 results are mixed and inconclusive.[5] Again, as a direct comparison with Berg et al. (1995), we also compared the reduction in entropy for the amounts invested between Sessions I and II. We have $\text{Entropy}_{\text{session I}} = 2.725$ and $\text{Entropy}_{\text{session II}} = 2.281$, indicating a reduction in entropy of 0.444. Bootstrapping showed that this entropy reduction of 0.444 is not significant at conventional levels.

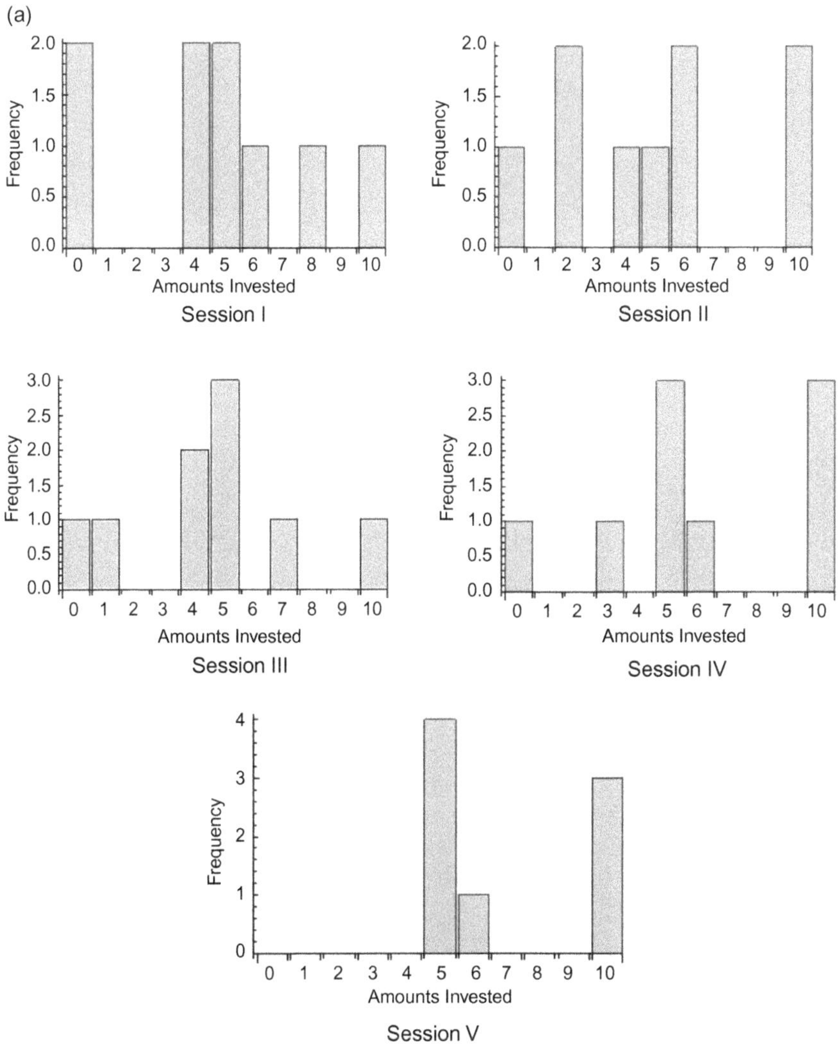


*Fig. 1.* Amounts Invested in Each of the Five Sessions: (a) Group 1 and (b) Group 2.

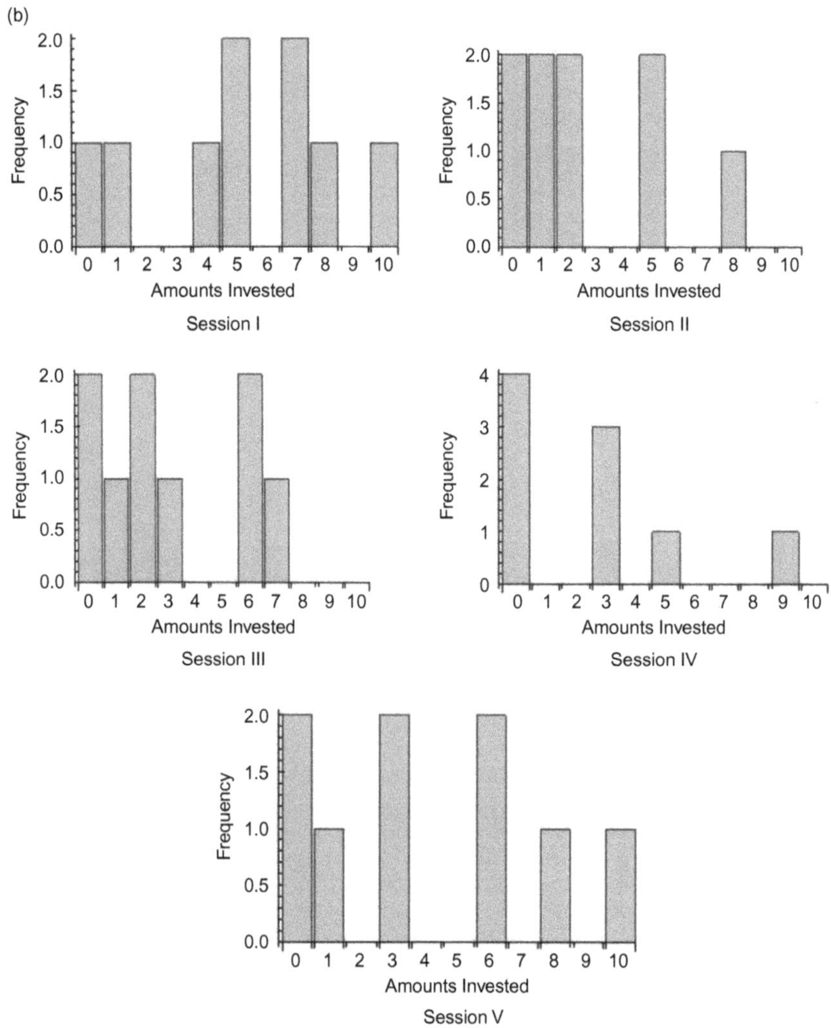


*Fig. 1.* (*Continued*)

*Steward Behavior*

H4 predicted that the introduction of financial history would reduce the entropy of amounts returned by stewards. Since the maximum amounts returned by stewards were constrained by the tripled amounts sent by

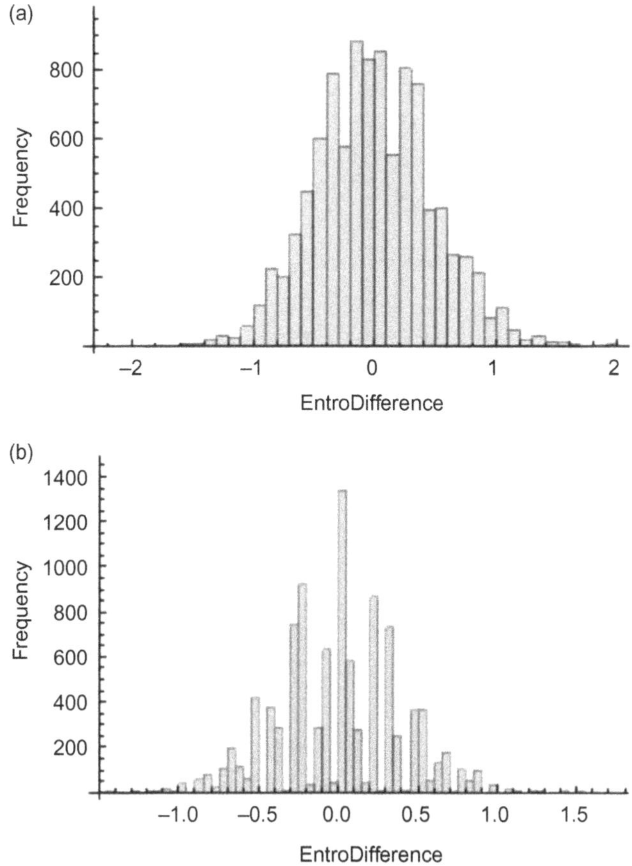


*Fig. 2.* Bootstrapped Distribution for Amounts Invested between Sessions I and V: (a) Group 1 and (b) Group 2.

investors, we analyzed the entropy in the distributions of the amounts returned by Stewards, converted into deciles of percentage of amounts received from Investors. Panel B of Table 3 and Panel B of Table 4 list the number of stewards who returned, by deciles of percentages of amounts received from investors and experimental session, for Group 1 and Group 2, respectively. For Group 1, the average amounts returned by stewards as a percentage of amounts received from Sessions I to V are 23.8%, 36.6%, 41.45%, 43.1%, and 41.25%, respectively. For Group 2, the average amounts returned by stewards as a percentage of amounts received from

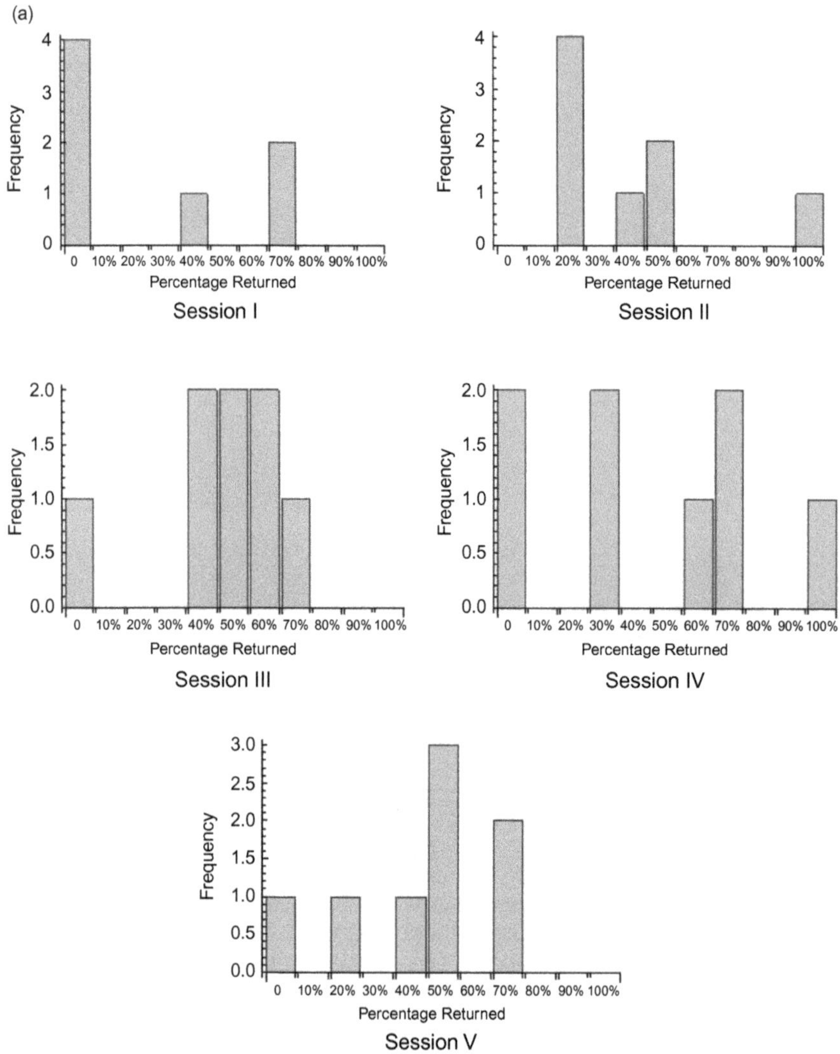


*Fig. 3.* Amounts Returned in Each of the Five Sessions, in Deciles of the Percentage of Amounts Received: (a) Group 1 and (b) Group 2.

(b)

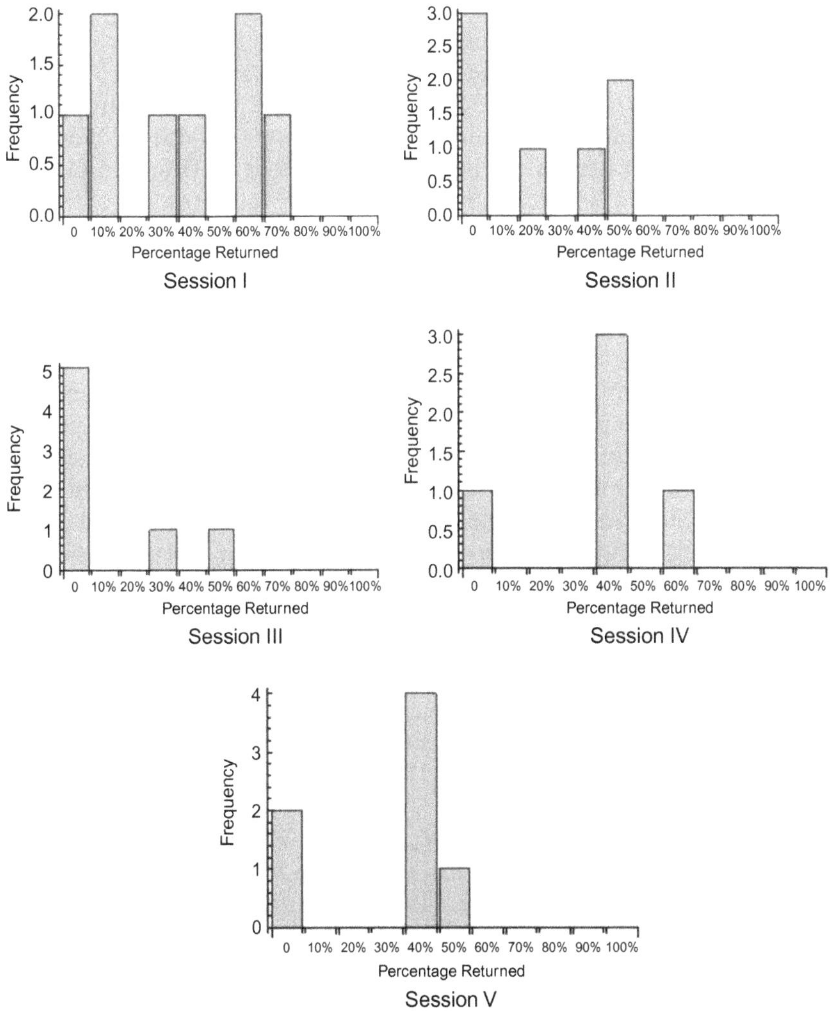


*Fig. 3.* (*Continued*)

Sessions I to V are 30.98%, 19.76%, 9.75%, 31.85%, and 25.39%, respectively. Fig. 3(a) (Fig. 3(b)) shows histograms of the Group 1 (Group 2) amounts returned in deciles of amounts received, for each of the five sessions. The Group 1 Steward distributions in Fig. 3(a) do not reveal a trend toward less variance across the five sessions. In contrast, the Group 2 Steward distributions in Fig. 3(b) reveal a trend toward less variance, though not in a socially beneficial direction, as the average percentage of amounts returned by stewards generally decreases across sessions.[6]

More formally, for Group 1, $\text{Entropy}_{\text{session I}} = 1.379$ and $\text{Entropy}_{\text{session V}} = 2.156$, indicating a trend toward more entropy (less order). We again used a bootstrapping technique to determine whether the observed Entropy difference of 0. 777 is statistically significant. The null hypothesis for this test is that the returned amounts are chosen randomly. Fig. 4(a) displays the bootstrapping results for Group 1. Based on this bootstrapped distribution, Session V has a significantly higher entropy compared to Session I (one-tailed $p < .05$).[7] This result is contrary to H4 and our entropy analysis of the Berg et al. (1995) Steward distributions. We conjecture that our result may be attributed to how Group 1 Session I Stewards exhibited more frequent zero amounts returned. The subsequent greater dispersion in the amounts returned by Group 1 Stewards in Sessions II–V resulted in higher entropy, but also indicates that a financial history yielded more trustworthy stewards. Consistent with this interpretation, Berg et al. (1995, p. 132) speculated (citing Coleman, 1990) that the provision of a financial history might enable Stewards to more easily identify with other investment game participants, and thus trigger internalized social norms of trustworthiness. As a direct comparison with Berg et al. (1995), we also compared the reduction in entropy for the amounts invested between Sessions I and II. We have $\text{Entropy}_{\text{session I}} = 1.379$ and $\text{Entropy}_{\text{session II}} = 1.75$, indicating an increase in entropy of 0.371. Bootstrapping showed that this entropy reduction of 0.371 is not significant at conventional levels.

In contrast, for Group 2, $\text{Entropy}_{\text{session I}} = 2.500$ and $\text{Entropy}_{\text{session V}} = 1.379$, indicating a trend toward less entropy (more order). We again used a bootstrapping technique to determine whether the observed Entropy difference of 1.121 is statistically significant. The null hypothesis for this test is that the returned amounts are chosen randomly. Fig. 4(b) displays the bootstrapping results for Group 2. Based on this bootstrapped distribution, Session V has a significantly lower entropy than Session I ($p < .01$, one-tailed).[8] These Group 2 results are consistent with H4 and our entropy analysis of the Berg et al. (1995) Steward distributions. In addition, when combined with the observation that the percentage of amounts returned by steward decreased as more financial history was revealed, it appears that the provision of a financial history enabled stewards to more easily identify whether other investment game participants were trustworthy or not.

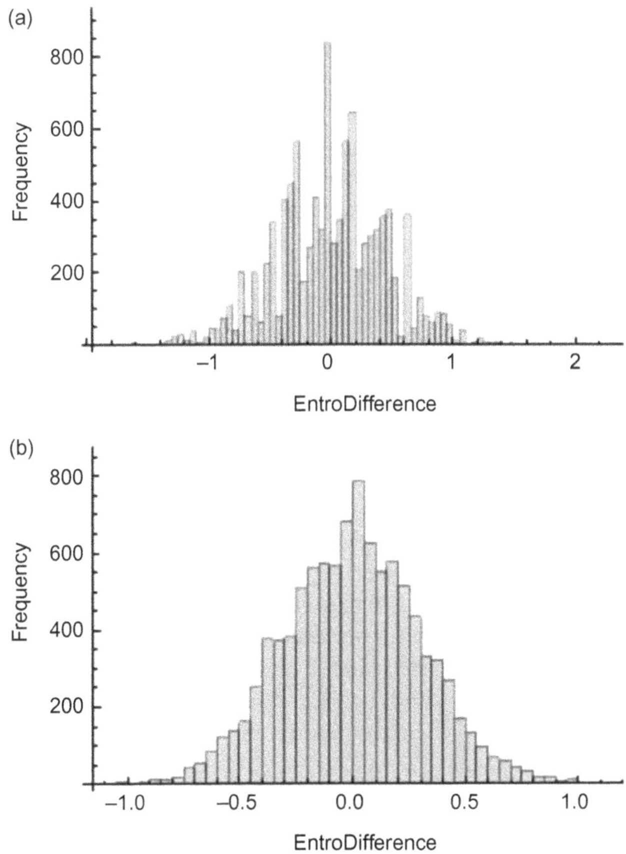


*Fig. 4.* Bootstrapped Distribution for Percentage Amounts Returned between Sessions I and V: (a) Group 1 and (b) Group 2.

Specifically, the financial history based on Group 2 Session I may have led Group 2 Session II stewards to infer a social norm whereby most participants are not expected to be trustworthy, and consequently more of them returned zero even when they received positive amounts from investors. Overall, however, the Group 1 and Group 2 results provide mixed support for H4's prediction that financial history would reduce the entropy in the amounts returned by Stewards.[9] As a direct comparison with Berg et al. (1995), we also compared the reduction in entropy for the amounts invested between Sessions I and II. We have $\text{Entropy}_{\text{session I}} = 2.500$ and

$\text{Entropy}_{\text{session II}} = 1.842$, indicating an increase in entropy of 0.658. Bootstrapping showed that this entropy reduction of 0.658 is marginally significant ($p < .1$, one-tailed).

*Joint Test of Investor/Steward Behavior*

We now examine the joint densities of Investor and Steward behavior to test whether the observed clustering is statistically significant, as predicted by H5. Since each session has the same sample size of 9, the effect of sample size on the entropy metric does not arise in our setting. The entropy formula is as follows:

$$-\sum_i \sum_j \frac{f_{i,j}}{N} \log_2 \frac{f_{i,j}}{N}$$

where $f_{ij}$ denotes the frequency in category "*i,j*." For example, "3,2" denotes 3 dollars invested and 2 returned by the steward and $N$ represents the number of sample observations. Under this metric for Group 1, $\text{Entropy}_{\text{session I}} = 2.948$ and $\text{Entropy}_{\text{session V}} = 2.750$. Our bootstrapped distribution indicates that the 0.198 reduction in entropy in the Group I joint Investor/Steward distributions from Session I to V is not significant (one-tailed $p > .4$). The results for Group 2 and a combination of Group 1 and Group 2 are qualitatively similar. As a direct comparison with Berg et al. (1995), we also compared the reduction in entropy for the amounts invested between our Sessions I and II. Again, the results are not statistically significant. Altogether, these results do not support H5, and are contrary to the marginally significant reduction in entropy we observed from Session I to Session II in the Berg et al. (1995) joint distributions (Table 5; Fig. 5).

## GENERAL DISCUSSION

We used the Berg et al. (1995) data to show that a report summarizing the financial transactions of a previous experimental session significantly reduced entropy in the amounts sent by Investors and returned by Stewards, and marginally reduced the entropy in the joint distribution of amounts invested and returned. We then sought to replicate these Berg et al. (1995) findings as well as extend their study by investigating whether financial histories will further reduce entropy over subsequent generations. More specifically, we investigated whether the provision of financial

***Table 5.*** Number of Investors in Group 1 and Group 2 Combined, Listed by Experimental Session and Amounts Sent and Returned.

Panel A. Number of investors in both groups, by experimental session and amount sent.

| Amounts Sent | Session I | Session II | Session III | Session IV | Session V |
|---|---|---|---|---|---|
| 0 | 1 | 3 | 3 | 5 | 2 |
| 1 | 1 | 2 | 2 | 0 | 1 |
| 2 | 0 | 4 | 2 | 0 | 0 |
| 3 | 0 | 0 | 1 | 4 | 2 |
| 4 | 1 | 1 | 2 | 0 | 0 |
| 5 | 2 | 3 | 3 | 4 | 4 |
| 6 | 0 | 2 | 2 | 1 | 3 |
| 7 | 2 | 0 | 2 | 0 | 0 |
| 8 | 1 | 1 | 0 | 0 | 1 |
| 9 | 0 | 0 | 0 | 1 | 0 |
| 10 | 1 | 2 | 1 | 3 | 4 |
| Entropy | 3.01307 | 2.86383 | 3.08605 | 2.3719 | 2.63129 |

Panel B. Number of stewards in Group 1 and Group 2 combined, listed by experimental session and amounts returned in deciles of percentage of amounts received from investors. Observations where investors sent $0 are excluded.

| Deciles of Percentage | Session I | Session II | Session III | Session IV | Session V |
|---|---|---|---|---|---|
| 0 | 5 | 3 | 6 | 3 | 3 |
| (0%, 10%] | 2 | 0 | 0 | 0 | 0 |
| (10%, 20%] | 0 | 5 | 0 | 0 | 1 |
| (20%, 30%] | 1 | 0 | 1 | 2 | 0 |
| (30%, 40%] | 2 | 2 | 2 | 3 | 5 |
| (40%, 50%] | 0 | 4 | 3 | 0 | 4 |
| (50%, 60%] | 2 | 0 | 2 | 2 | 0 |
| (60%, 70%] | 3 | 0 | 1 | 2 | 2 |
| (70%, 80%] | 0 | 0 | 0 | 0 | 0 |
| (80%, 90%] | 0 | 0 | 0 | 0 | 0 |
| (90%, 100%] | 0 | 1 | 0 | 1 | 0 |
| Entropy | 2.41592 | 2.14926 | 2.28925 | 2.50738 | 2.14926 |

transaction history from preceding generations will enable successive generations of a laboratory society to increasingly organize themselves over time. We operationally defined an increase in organization as a reduction in entropy of the distributions of amounts invested and returned in an investment game. To test the predictive power of our main hypothesis that financial histories over successive generations will enable the emergence of increasing societal order, we gathered new data from two groups (Group 1 and Group 2) over five societal generations.

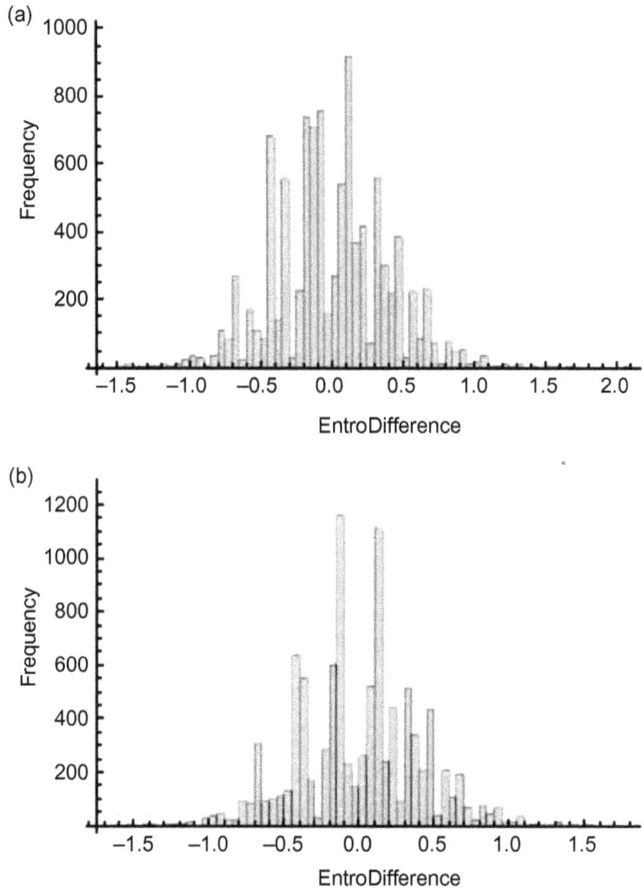


*Fig. 5.* Bootstrapped Distribution for the Joint Behavior between Sessions I and V: (a) Group 1 and (b) Group 2.

If we assume that all agents are myopically self-interested and believe that all other agents are similarly self-interested, then each Investor and Steward should "take the money and run," leaving no role for financial history in the evolution of society. In order for financial history to play a role in our experimental setting, the assumption of myopic self-interest needs to be relaxed. For example, if each Investor is myopically self-interested but believes that his or her partner might not be, then a financial history of past Investor/Steward behavior can tell Investors something about the likely amount returned for each amount invested. Thus, financial histories might influence the amounts invested, and if Stewards are not myopically self-interested, financial histories might also give Stewards a basis to infer and mimic expected or acceptable behavior.

Consistent with the prior investment game literature and our H1, our bootstrap analyses indicate that the distributions of amounts sent by Investors in all five sessions for both Group 1 and Group 2 differed significantly from zero. Similarly, the distributions of amounts returned by Stewards generally supported our H2 but the results were more mixed. Specifically, the distributions of amounts returned by Stewards were significantly greater than zero for four of five Group 1 Sessions and three of five Group 2 sessions.

Overall, however, the pattern of results across our study and the Berg et al. (1995) study indicates that the impact of financial history on entropy reduction is not as robust as initially thought. More specifically, consistent with our H3, our bootstrap analyses indicate that financial history generally reduced the entropy in the amounts sent by Investors across the five sessions, with stronger results for Group 1 than for Group 2. In contrast, we found mixed support for a reduction in entropy in the amounts returned by Stewards (H4). Specifically, the entropy in the Group 2 Session V amounts returned by Stewards was significantly lower than the entropy in the Group 2 Session I amounts, as predicted by our H4, but the entropy in the Group 1 Session V amounts returned by Stewards was higher rather than lower than the entropy in the Group 1 Session I amounts. Finally, our analyses did not support our H5 prediction that financial history would reduce entropy in the joint Investor/Steward distributions. The combined entropy results for H3–H5 yielded a similar pattern. Clearly, more research is needed to document the potential mediating effects of financial history across a variety of conditions. Below, we discuss several opportunities for future research.

First, along with Berg et al. (1995), we conjectured that the provision of a financial history might enable stewards to more easily identify whether other Investment Game stewards are trustworthy. This suggests that the impact of financial history may depend on initial conditions and is path dependent. For example, if financial histories indicate that past stewards tended to be trustworthy, current and future stewards might internalize social norms of trustworthiness and consequently tend to act in a more trustworthy manner. Likewise, current and future investors might internalize social norms of trusting behavior and consequently be willing to invest more. The opposite Investor and Steward patterns may emerge if financial histories indicate that stewards tend to return zero or very low amounts. A larger set of experiments would be required to formally test how initial conditions and other path dependencies might interact with financial histories.

Second, entropy reduction is only one of several possible social outcomes that could be mediated by financial histories. Indeed, as we observed in our data, it is possible for financial histories to yield more disperse but also more trustworthy Stewardship distributions, or instead yield less diverse but also less trustworthy Stewardship distributions. For example, an emergent increase in Stewardship trustworthiness might in turn yield more trusting Investors over subsequent generations. These iterative effects could ultimately yield greater overall social wealth, despite higher entropy in the amounts returned by Stewards. On the other hand, if Investment Game participants tend to distrust each other in the beginning, financial histories might shape participants' perceptions of other people and thereby augment their distrust. This would in turn yield lower entropy but lower total social wealth as the investors send and stewards return lower amounts. Again, a larger set of experiments would be required to examine these potential effects.

Third, variations of the Investment Game could be used to test the incremental effects of financial history when Investors and Stewards participate in multiple sessions and are allowed to communicate with one another (cf. Charness, Duh, & Yang, 2011; Chaudhuri et al., 2009), or when social distance between the experimenter and subjects is varied (cf. Cox & Deck, 2005; who found more positively reciprocal behavior in the Trust Game when the experimenter had personal knowledge of subjects' behavior).

Fourth, Ortmann et al. (2000) found that the amounts invested by investors in the Investment Game were robust to whether the investors completed a questionnaire to prompt their strategic reasoning about stewards' likely behavior, and whether the social history was presented in table form only or in tables plus graphs. Future research could test whether such modifications will influence the amounts returned by stewards and stewards' strategic reasoning about investor behavior. But more generally, the Johnson and Mislin (2011) meta-analysis of investment game studies found mixed results from prior studies that engaged subjects to think strategically through the various behavioral implications of each possible outcome. This suggests that the effect of conscious strategies may be highly contingent on other factors that remain to be identified and experimentally controlled.

Finally, although the reporting of financial histories can be a conscious strategy that societal agents use to structure their economic and social environments, they also can have unintended effects on organizations and societies, as has been found in other measurement systems (Burchell, Clubb, Hopwood, Hughes, & Nahapiet, 1980; Cyert & March, 1992; Thompson, 1967). While the effects of financial histories might in some circumstances be mediated by the conscious strategies of individual agents (cf. Berg et al., 1995, p. 134), financial histories also might structure the behavior of subsequent generations without necessarily requiring agents to be consciously aware of these effects (cf. Giddens, 1984). More research is needed to ascertain the extent to which the effect of financial history is mediated by conscious versus unconscious individual perceptions and decision strategies.

## NOTES

1. Ortmann et al.'s (2000) study of the Investment Game reported that the amounts invested were robust to whether the Investors completed a questionnaire designed to prompt strategic reasoning, and to changes in how the social history information was presented to subjects (e.g., in tables only as in Berg et al. (1995) or in tables accompanied by graphs that showed the proportional amounts returned by stewards).

2. Specifically, when the probability of heads is 0.75, the entropy is calculated as $-0.75\log_a(0.75) - 0.25\log_a(0.25)$ where $a$ is the base of the logarithm. The result is approximately 0.81 when the binary logarithm (log base 2) is used. If we had instead calculated entropy with the natural logarithm (log base e), the result would have been approximately 0.56. The choice of base affects the absolute value of the entropy measure but it does not affect the comparison of entropy measures when all of the measures are calculated using the same base. All of our entropy calculations use the binary logarithm.

3. Note that variance captures the same construct as entropy in a normal distribution. For non-normal distributions, however, entropy has been shown to be a better measure of uncertainty than variance. Nevertheless, we also conducted a variance analysis with qualitatively similar results.

4. Specifically, in the No History condition there are at most 32 distinct observations. This means the maximum level of entropy will be $-\log_2(1/32) = 5.0$. In contrast, if there were only 28 observations, the maximum entropy would be $-\log_2(1/28) = 4.8$.

5. Combining data from both Group 1 and Group 2, $\text{Entropy}_{\text{session I}} = 3.013$ and $\text{Entropy}_{\text{session V}} = 2.631$, indicating a marginally significant ($p = .10$, one-tailed) reduction in entropy of 0.382.

6. Note that while a scenario where all stewards return 100% or all stewards return 0% will generate the least amount of entropy, the former is socially beneficial whereas the latter is not.

7. Pairwise entropy differences between Group 1 Sessions II and V are not significant at conventional levels.

8. Pairwise entropy differences between Group 1 Sessions II and V are not significant at conventional levels.

9. Combining data from both Group 1 and Group 2, $\text{Entropy}_{\text{session I}} = 2.416$ and $\text{Entropy}_{\text{session V}} = 2.149$. Session V has lower entropy, but a bootstrapping test indicates that the difference of 0.267 is not significant at conventional levels.

## ACKNOWLEDGMENTS

The authors thank Cary Deck (co-editor) and an anonymous reviewer for their helpful comments and suggestions, and thank John Dickhaut and Judy Rayburn for their ideas, inspiration, and encouragement in the early stages of this research. Research funding from the Merage School of Business (University of California – Irvine), the Department of Accounting at the Carlson School of Management (University of Minnesota – Twin Cities), and the University of St. Thomas Opus College of Business is gratefully acknowledged.

## REFERENCES

Berg, J., Dickhaut, J., & McCabe, K. (1995). Trust, reciprocity, and social history. *Games and Economic Behavior*, *10*, 122–142. doi:10.1006/game.1995.1027

Burchell, S., Clubb, C., Hopwood, A. G., Hughes, J., & Nahapiet, J. (1980). The roles of accounting in organizations and society. *Accounting, Organizations and Society*, *5*, 5–27. doi:10.1016/0361-3682(80)90017-3

Camerer, C. F., & Fehr, E. (2002). Measuring social norms and preferences using experimental games: A guide for social scientists. In J. Henrich, R. Boyd, S. Bowles, C. Camerer, E. Fehr & H. Gintis (Eds.), *Experimental and ethnographic evidence from 15 small-scale societies* (pp. 55–95). New York, NY: Oxford University Press.

Charness, G., Du, N., & Yang, C. (2011). Trust and trustworthiness reputations in an investment game. *Games and Economic Behavior*, *72*, 361–375. doi:10.1016/j.geb.2010.09.002

Chaudhuri, A., Schotter, A., & Sopher, B. (2009). Talking ourselves to efficiency: Coordination in inter-generational minimum effort games with private, almost common and common knowledge of advice. *The Economic Journal*, *119*, 91–122. Retrieved from http://www.jstor.org/stable/20485296

Coleman, J. (1990). *Foundations of social choice theory*. Cambridge, MA: Harvard University Press.

Cover, T. M., & Thomas, J. A. (1991). *Elements of information theory*. New York, NY: Wiley-Interscience.

Cox, J. C. (2004). How to identify trust and reciprocity. *Games and Economic Behavior*, *46*, 260–281. doi:10.1016/S0899-8256(03)00119-2

Cox, J. C., & Deck, C. A. (2005). On the nature of reciprocal motives. *Economic Inquiry*, *43*(3), 623–635. doi:10.1093/ei/cbi043

Cyert, R., & March, J. (1992). *A behavioral theory of the firm* (2nd ed.). Oxford: Basil Blackwell.

Dickhaut, J., & McCabe, K. A. (1997). The behavioral foundations of stewardship accounting and a proposed program of research: What is accountability? *Behavioral Research in Accounting*, *9*, 60–87. Retrieved from http://search.ebscohost.com/login.aspx?direct=true&db=bth&AN=9708106809&site=ehost-live

Giddens, A. (1984). *The constitution of society*. Cambridge, MA: Polity Press.

Habermas, J. (1979). *Communication and the evolution of society* (T. McCarthy, Trans.). Boston, MA: Beacon Press.

Johnson, N. D., & Mislin, A. A. (2011). Trust games: A meta-analysis. *Journal of Economic Psychology*, *32*, 865–889. doi:10.1016/j.joep.2011.05.007

Kramer, R. M. (1999). Trust and distrust in organizations: Emerging perspectives, enduring questions. *Annual Review of Psychology*, *50*, 569–598. doi:10.1146/annurev.psych.50.1.569

Latour, B. (1993). *We have never been modern*. Harvard, MA: Harvard University Press.

Leydesdorff, L. (2001). *A sociological theory of communication: The self-organization of the knowledge-based society*. uPublish.com. Boca Raton, FL: Universal Publishers. Retrieved from http://www.upublish.com/books/leydesdorff.htm

Luhmann, N. (1986). The autopoiesis of social systems. In F. Geyer & J. van der Zouwen (Eds.), *Sociocybernetic paradoxes: Observation, control and evolution of self-steering systems* (pp. 172–192). London: Sage.

Lunawat, R. (2013a). An experimental investigation of reputation effects of disclosure in an investment/trust game. *Journal of Economic Behavior & Organization*, *94*, 130–144. doi:10.1016/j.jebo.2013.07.018

Lunawat, R. (2013b). The role of information in building reputation in an investment/trust game. *European Accounting Review*, *22*(3), 513–532. Retrieved from http://search.ebscohost.com/login.aspx?direct=true&db=bth&AN=89600534&site=ehost-live

Mackowiak, B., & Wiederholt, M. (2009). Optimal sticky prices under rational inattention. *American Economic Review*, *99*(3), 769–803.

Ortmann, A., Fitzgerald, J., & Boeing, C. (2000). Trust, reciprocity, and social history: A re-examination. *Experimental Economics*, *3*, 81–100. doi:10.1023/A:1009946125005

Shannon, C. E. (1948). A mathematical theory of communication. *The Bell System Technical Journal*, *27*(July), 379–423. doi:10.1145/584091.584093

Sims, C. A. (1998). Stickiness. *Carnegie-Rochester Conference Series on Public Policy*, *49*(1), 317–356.

Sims, C. A. (2003). Implications of rational inattention. *Journal of Monetary Economics*, *50*(3), 665–690. doi:10.1016/S0304-3932(03)00029-1

Sims, C. A. (2005). *Rational inattention: A research agenda*. Deutsche bundesbank research centre discussion paper series 1: Economic studies, 2005, 34.

Thompson, J. D. (1967). *Organizations in action: Social science bases of administrative theory*. New York, NY: McGraw-Hill.

Yang, M. (2015). Coordination with flexible information acquisition. *Journal of Economic Theory*, *158*, 721–738. doi:10.1016/j.jet.2014.11.017

# APPENDIX

*Instructions for Room A*

You have been asked to participate in a decision-making study. I will read these instructions out loud. Please do not talk among yourselves. If you have any questions, please raise your hand. I will then answer your questions individually.

*Overview*
In this experiment each of you will be paired with a different participant who is in another room. You will not be told who these people are either during or after the experiment. **This is Room A.** Other participants are in Room B. You will notice that there are other people in the same room with you who are also participating in this experiment. You will not be paired with any of these people.

One participant in Room A will be called Monitor A, and one participant in Room B will be called Monitor B. Monitor A and Monitor B will be randomly chosen before the experiment begins. The monitors will be in charge of the envelopes as explained below. The monitors also will verify that the instructions have been followed. Monitors will be given $15 at the end of the experiment.

Each participant in Room B who is not a monitor will be given $10 for showing up on time and participating until the end of the experiment. Each participant in Room A will receive $10 in an envelope. Each Room A participant will then have the opportunity to send some, all, or none of the $10 to a participant in Room B. Each dollar sent to Room B will be tripled. For example, if a participant in Room A sends an envelope containing $2, he or she will keep the remaining $8 and the envelope will contain $6 when it reaches Room B. If a participant in Room A sends an envelope containing $9, he or she will keep the remaining $1 and the envelope will contain $27 when it reaches Room B. The participant in Room B will then decide how much money to send back to the participant in Room A.

(For the financial history treatment we added the following paragraph: Each of you has received a report summarizing the decisions of subjects who participated in a previous experiment. Please check the last page of the instructions to be sure you have this sheet.)

*Confidentiality*
This experiment is designed so that no one, including the experimenters and monitors, will know the decisions of any individual participants in Room A or Room B. Since your decision is private, we ask that you do not tell anyone your decision either during or after the experiment.

*How the Experiment is Run*
*Room A Participant Decisions.* The experiment is conducted as follows. Ten large unmarked envelopes have been placed in a box in Room A. Each of these envelopes contains 10 one dollar bills, a smaller inner envelope, and a key in a sealed envelope marked "KEY." The inner envelope and key are both marked with the same letter of the alphabet. The Room A monitor will randomly select and hand each participant in Room A an unmarked envelope from the box. Each Room A participant will perform the following steps:

1. Privately open the unmarked envelope. The participant who opens the envelope will be the only person who knows which letter of the alphabet was in the envelope.
2. Leave the envelope marked "KEY" unopened until instructed to open it later in the experiment.
3. Privately decide how many dollar bills to put into the inner envelope, and then put those dollar bills in the inner envelope.
4. Privately pocket any remaining dollar bills and pocket the unopened envelope marked "KEY."
5. Privately put the inner envelope back inside the large unmarked envelope.
6. Raise his or her hand. When Monitor A comes by, return the unmarked envelope to the box marked "return the envelopes here."

*Recording and Processing of Room A Envelopes.* As soon as all Room A envelopes have been put into the return box, Monitor A will transport the box to a recorder who is sitting in the hallway. With Monitor A observing, the recorder will perform the following steps:

1. Take all unmarked envelopes out of the box.
2. Take an inner envelope out of an unmarked envelope, one at a time.
3. On a blank sheet of paper, record the letter that is printed on the envelope and the amount of money in the envelope.

4. Triple the amount of money in the inner envelope, place the inner envelope back into the unmarked outer envelope, and place the unmarked outer envelope back in the box.
5. After all of the envelopes have been processed in this manner, the recorder will signal Monitor B to come to the recorder's desk. Once Monitor B has arrived, Monitor A will be asked to return to Room A.

*Room B Participant Decisions.* Monitor B will carry the box of envelopes to Room B. Monitor B then will randomly select and hand each participant in Room B an unmarked envelope from the box. Each Room B participant will perform the following steps:

1. Decide how many dollar bills to leave in the inner envelope.
2. Pocket any remaining dollar bills.
3. Place the inner envelope in the unmarked outer envelope.
4. Raise his or her hand. When Monitor B comes by, return the outer envelope to the box marked "return envelopes here."
5. Gather his or her belongings, receive his or her $10 compensation for participating in the experiment, and be asked to leave the building.

*Recording and Processing of Room B Envelopes.* After all envelopes in Room B are returned to the box marked "return the envelopes here" and all Room B participants have been asked to leave the building, Monitor B will transport the box to the recorder in the hallway. With Monitor B observing, the recorder will perform the following steps:

1. Take all unmarked envelopes out of the box.
2. Take an inner envelope out of an unmarked envelope, one at a time.
3. Record on a blank sheet of paper the letter on the envelope and the amount of money in the envelope.
4. Place the inner envelope in the box.
5. After all of the envelopes have been processed in this manner, the recorder will then signal Monitor A to come to the recorder's desk.

Once Monitor A has arrived, Monitor B will return to Room B. When Monitor A arrives, Monitor A and the recorder will carry the box of envelopes to Room C. Room C contains locking boxes with identifying letters. The letters correspond to the letters on the inner envelopes. While the recorder observes, Monitor A will place each inner envelope in the box with the corresponding letter. All of the boxes will then be locked. The recorder will then return to the hallway and Monitor A will return to Room A.

*Room A Participants Privately Open the Returned Envelopes.* Monitor A will then point to one participant in Room A at a time to proceed to Room C. When called to go to Room C, each Room A participant will perform the following steps:

1. Gather all of his or her belongings since he or she will be asked to leave the building when done.
2. Enter Room C alone.
3. Open the envelope marked "KEY." Inside this envelope is a lettered key which will open the locked box with the corresponding letter. The inner envelope in the box is the same envelope that Room A participant started with.
4. Go to the appropriate locked box, open it, take out the envelope, and remove the money.
5. Return the empty envelope to the box.
6. Lock the box.
7. Return the key to the envelope marked "KEY."
8. Drop the envelope marked "KEY" in the box just outside the door in the hallway and be asked to leave the building.

*Conclusion of Experiment.* After everyone in Room A has left, the experiment is over and the two monitors will be paid $15 for their participation.

*Note about Dollar Bills.* We will use e-dollars for this experiment. At the end of the experiment, your e-dollars will be converted into real dollars at a faceless teller. The conversion rate is one dollar for one e-dollar bill.

## *Instructions for Room B*

You have been asked to participate in a decision-making study. I will read these instructions out loud. Please do not talk among yourselves. If you have any questions, please raise your hand. I will then answer your questions individually.

*Overview*
In this experiment each of you will be paired with a different participant who is in another room. You will not be told who these people are either during or after the experiment. **This is Room B.** Other participants are in Room A. You will notice that there are other people in the same room with you who are also participating in this experiment. You will not be paired with any of these people.

One participant in Room A will be called Monitor A, and one participant in Room B will be called Monitor B. Monitor A and Monitor B will be randomly chosen before the experiment begins. The monitors will be in charge of the envelopes as explained below. The monitors also will verify that the instructions have been followed. Monitors will be given $15 at the end of the experiment.

Each participant in Room B who is not a monitor will be given $10 for showing up on time and participating until the end of the experiment. Each participant in Room A will receive $10 in an envelope. Each Room A participant will then have the opportunity to send some, all, or none of the $10 to a participant in Room B. Each dollar sent to Room B will be tripled. For example, if a participant in Room A sends an envelope containing $2, he or she will keep the remaining $8 and the envelope will contain $6 when it reaches Room B. If a participant in Room A sends an envelope containing $9, he or she will keep the remaining $1 and the envelope will contain $27 when it reaches Room B. The participant in Room B will then decide how much money to send back to the participant in Room A.

(For the financial history treatment we added the following paragraph: Each of you has received a report summarizing the decisions of subjects who participated in a previous experiment. Please check the last page of the instructions to be sure you have this sheet.)

*Confidentiality*
This experiment is designed so that no one, including the experimenters and monitors, will know the decisions of any individual participants in Room A or Room B. Since your decision is private, we ask that you do not tell anyone your decision either during or after the experiment.

*How the Experiment is Run*
*Room A Participant Decisions*. The experiment is conducted as follows. Twelve large unmarked envelopes have been placed in a box in Room A. Each of these envelopes contains 10 one dollar bills, a smaller inner envelope, and a key in a sealed envelope marked "KEY." The inner envelope

and key are both marked with the same letter of the alphabet. The Room A monitor will randomly select and hand each participant in Room A an unmarked envelope from the box. Each Room A participant will perform the following steps:

1. Privately open the unmarked envelope. The participant who opens the envelope will be the only person who knows which letter of the alphabet was in the envelope.
2. Leave the envelope marked "KEY" unopened until instructed to open it later in the experiment.
3. Privately decide how many dollar bills to put into the inner envelope, and then put those dollar bills in the inner envelope.
4. Privately pocket any remaining dollar bills and pocket the unopened envelope marked "KEY."
5. Privately put the inner envelope back inside the large unmarked envelope.
6. Raise his or her hand. When Monitor A comes by, return the unmarked envelope to the box marked "return the envelopes here."

*Recording and Processing of Room A Envelopes.* As soon as all Room A envelopes have been put into the return box, Monitor A will transport the box to a recorder who is sitting in the hallway. With Monitor A observing, the recorder will perform the following steps:

1. Take all unmarked envelopes out of the box.
2. Take an inner envelope out of an unmarked envelope, one at a time.
3. On a blank sheet of paper, record the letter that is printed on the envelope and the amount of money in the envelope.
4. Triple the amount of money in the inner envelope, place the inner envelope back into the unmarked outer envelope, and place the unmarked outer envelope back in the box.
5. After all of the envelopes have been processed in this manner, the recorder will signal Monitor B to come to the recorder's desk. Once Monitor B has arrived, Monitor A will be asked to return to Room A.

*Room B Participant Decisions.* Monitor B will carry the box of envelopes to Room B. Monitor B then will randomly select and hand each participant in Room B an unmarked envelope from the box. Each Room B participant will perform the following steps:

1. Decide how many dollar bills to leave in the inner envelope.
2. Pocket any remaining dollar bills.

3. Place the inner envelope in the unmarked outer envelope.
4. Raise his or her hand. When Monitor B comes by, return the outer envelope to the box marked "return envelopes here."
5. Gather his or her belongings, receive his or her $10 compensation for participating in the experiment, and be asked to leave the building.

*Recording and Processing of Room B Envelopes.* After all envelopes in Room B are returned to the box marked "return the envelopes here" and all Room B participants have been asked to leave the building, Monitor B will transport the box to the recorder in the hallway. With Monitor B observing, the recorder will perform the following steps:

1. Take all unmarked envelopes out of the box.
2. Take an inner envelope out of an unmarked envelope, one at a time.
3. Record on a blank sheet of paper the letter on the envelope and the amount of money in the envelope.
4. Place the inner envelope in the box.
5. After all of the envelopes have been processed in this manner, the recorder will then signal Monitor A to come to the recorder's desk.

Once Monitor A has arrived, Monitor B will return to Room B. When Monitor A arrives, Monitor A and the recorder will carry the box of envelopes to room C. Room C contains locking boxes with identifying letters. The letters correspond to the letters on the inner envelopes. While the recorder observes, Monitor A will place each inner envelope in the box with the corresponding letter. All of the boxes will then be locked. The recorder will then return to the hallway and Monitor A will return to Room A.

*Room A Participants Privately Open the Returned Envelopes.* Monitor A will then point to one participant in Room A at a time to proceed to Room C. When called to go to Room C, each Room A participant will perform the following steps:

1. Gather all of his or her belongings since he or she will be asked to leave the building when done.
2. Enter Room C alone.
3. Open the envelope marked "KEY." Inside this envelope is a lettered key which will open the locked box with the corresponding letter. The inner envelope in the box is the same envelope that Room A participant started with.

4. Go to the appropriate locked box, open it, take out the envelope, and remove the money.
5. Return the empty envelope to the box.
6. Lock the box.
7. Return the key to the envelope marked "KEY."
8. Drop the envelope marked "KEY" in the box just outside the door in the hallway, receive his or her $10 compensation for participating in the experiment and be asked to leave the building.

*Conclusion of Experiment.* After everyone in Room A has left, the experiment is over and the two monitors will be paid $15 for their participation.

*Note about Dollar Bills.* We will use e-dollars for this experiment. At the end of the experiment, your e-dollars will be converted into real dollars at a faceless teller. The conversion rate is one dollar for one e-dollar.

# REVISITING THE EFFECT OF VOTER ISOLATION

Alexander G. James and Jason F. Shogren

## ABSTRACT

*Does social-isolation deflate observed preferences for public goods? Using a voting referendum elicitation mechanism, List et al. (2004) document a 30% reduction in affirmative voting when votes are privately, rather than publicly, cast for a public good. We replicate this work and then add to the exercise by examining the role that group size plays in generating social-pressure bias – an extension we motivate with a structural model. Having replicated to the extent possible the List et al.'s consequential private and public treatments, we find that when group size is large (N = 60), social isolation reduces affirmative votes by roughly 30%, a result that does not carry over to the small-group (N = 30) session.*

**Keywords:** Voting referendum; social isolation; preference revelation

**JEL Classifications:** C9; H4; Q5

**Replication in Experimental Economics**
**Research in Experimental Economics, Volume 18, 137–152**

**ISSN: 0193-2306/doi:10.1108/S0193-230620150000018005**

## INTRODUCTION

Economists have long promoted experimental methods as a tool to examine the behavioral underpinnings of alternative preference elicitation methods before they are implemented in stated preference surveys or field auctions (see, e.g., Cummings, Brookshire, & Schulze, 1986; Coursey & Schulze, 1986; Shogren, 2004). One popular elicitation method is the referendum voting procedure, in which nonmarket values are elicited based on the number of "yay" and "nay" votes to pay for the provision of a costly public good (see Carson & Groves, 2007). Given the discrete choice-style referendum is a collective mechanism, some economists working on nonmarket valuation issues have wondered whether nonmarket behavioral factors like social pressure or social isolation might affect how people choose to vote (see Berrens, Bohara, & Kerkvliet, 1997; Shogren & Taylor, 2008; Swim et al., 2009). The concern arose in part because of the substantial experimental literature that exists suggests social isolation tends to decrease a person's willingness to contribute/donate to costly public goods (see, e.g., Andreoni & Petrie, 2004; Bohnet & Frey, 1999; Rege & Telle, 2004).

Examining the robustness of this phenomenon, List, Berrens, Bohara, and Kerkvliet (2004) find that social isolation does affect referendum voting. They find that social isolation significantly decreases the probability that people will vote "yes" in a referendum for a costly public good (i.e., donations to a proposed on-campus Center on Environmental Policy Analysis). They conclude by questioning the validity of stated preference methods which did not control for social isolation: "We believe our results are fundamental to understanding the received evidence for mechanism design in [stated preference methods], as they serve to highlight the notion that utilitarian elements and strategic reciprocity ... are confounded in interpreting signals of value, and that potential biases introduced may be much larger than many expect (p. 742)." If their findings are as robust and large as they fear, future work in nonmarket valuation might need to control for the specific social context underlying the study (also see Fischhoff, 2005).

Herein, we design a referendum valuation experiment to replicate and test the robustness of whether social isolation affects referendum voting behavior for a costly public good. We first present an analytical model that suggests the referendum is particularly susceptible to social-pressure bias because within a referendum, the voter faces only a probability of being a decisive voter. This creates an incentive for a person to vote "yes" because he or she can gain social approval without risking any influence on the outcome of the referendum. Our model goes further though – it suggests that social pressure should decrease as the group size decreases. Our experimental data are consistent with these two predictions. Consistent with the observations in List et al., we find that social isolation decreased the frequency of "yay" votes in our larger group ($n=60$). In addition, we observe that social isolation had no significant effect on voting behavior within the relatively smaller group ($n=30$), as suggested by the motivating theory.

## A MOTIVATING FRAMEWORK

To motivate our replication experiment we consider a representative person who receives utility from the consumption of a composite good, $c_i$, total contributions to a public good, $A$, and by being perceived as valuing a public good by an audience.[1] He or she makes a decision from within a group and the audience is the voting bloc. Each person votes either "yes" or "no" for every member of the bloc to donate $t$ dollars toward a public good. If the majority (more than 50%) votes "yes," everyone in the bloc must donate. If the majority votes "no," no one in the bloc donates. Assume with probability $q$ the vote of person $i$ is revealed to the audience and with probability $p$ person $i$ is a tie-breaker, or a "decisive voter." The revelation of person $i$'s vote informs the audience's perception of person $i$'s true preference for the public good.

The representative person's optimization problem is

$$\max_{j\in 0,1} U_{i,VR} = (1-j)[p(c_i+t)+qhE(\nu_i|j=0)]+j[p(c_i+\nu_i g(tN))$$

$$+qhE(\nu_i|j=1)]+(1-p)\bar{x} \qquad (1)$$

by voting either "yes" ($j=1$) or "no" ($j=0$). Let $N$ be the total number of people in the voting bloc and $\bar{x}$ is the exogenously determined level of utility earned when the voter is indecisive. Let $g$ be a function that maps the level of the public good into a person's utility function. Assume $g' > 0$. Person $i$'s preference for the audience's perception of their preference for the public good is given by $h$ and $\nu_i$ is person $i$'s true preference for the public good.

A person's true preference for the public good, $\nu_i$, is one of two types: high or low. By voting "yes" for the public good, a person signals their preference for the public good is the high type. The opposite signal is sent by voting "no." Let the audience's expectation of $\nu_i$, given person $i$ votes "yes" in a voting referendum be given by $E(\nu_i|j=1)=\overline{\Omega}$ and let the audience's expectation of $\nu_i$ given person $i$ votes "no" in a voting referendum be given by $E(\nu_i|j=0)=\underline{\Omega}$.

Person $i$ is the decisive voter in the event his or her vote is a tie-breaker. Following Beck (1975), we define the probability of being a decisive voter as the probability the voting behavior of $N-1$ other voters results in a tie, in which $N$ is an odd number. Assuming that voter $i$ believes voter $k$ will vote "yes" with probability $\overline{p}$ (before any votes are cast) and that $N$ is large, the probability that voter $i$ is the decisive voter (the probability that the voting behavior of $N-1$ people results in a tie) is

$$p(\overline{p},N)=(4\overline{p}(1-\overline{p}))^{(N-1)/2}\left(\frac{2}{\pi(N-1)}\right)^{1/2} \tag{2}$$

Taking the derivative of Eq. (2) with respect to $N$ indicates that the probability of being the decisive voter is decreasing in the size of the voting bloc:

$$\frac{\partial p}{\partial N}=p\left(\ln(4\overline{p}(1-\overline{p}))-\frac{1}{2(N-1)}\right)<0 \tag{3}$$

for $0\leq\overline{p}\leq 1$. For example, suppose $\overline{p}=1/2$ and $N$ is 11. The probability a particular voter is decisive is 25%. If $N=61$, the probability of being a decisive voter drops to 10%.

Because a person must vote either "yes" or "no," he or she votes "yes" if the net utility of doing so is positive and "no" otherwise. The utility earned from voting "yes" less the utility earned from voting "no" is given by:

$$p(\nu_i g(tN)-t)+qh(\overline{\Omega}-\underline{\Omega}) \tag{4}$$

Note if voters are in total social isolation, $q=0$. In this extreme case, a person votes "yes" if they prefer to collectively give $tN$ to the public good than to privately consume $t$. A person's voting behavior is unaffected by the probability, $p$, of being the decisive voter. In the absence of social isolation,

$q>0$ and a person votes "yes" if Eq. (4) is positive. Now the probability of being the decisive voter plays a critical role. For a sufficiently small probability, $p$, of being decisive, even a weak preference for social approval can switch a "no" vote to a "yes" one. This result is specific to a referendum. It is not obvious that this bias would carry over to alternative elicitation mechanisms such as an open-ended questionnaire of dichotomous choice mechanism in which subjects face unitary probabilities of decisiveness. In sum, we expect social isolation to trigger more "yes" votes – particularly, when group size is high.

## EXPERIMENTAL DESIGN

To test the hypothesis that larger group sizes increase the number of "yes" votes, we use a $2\times2$ experimental design in which social isolation and group size are the treatments. We replicate List et al.'s public and private referendum format as close as possible and extend their analysis to consider alternative group sizes.[2]

Table 1 summarizes our experiment relative to List et al. We similarly asked undergraduate business school students to vote according to their preference for a public good. We use similar sample sizes but observe

***Table 1.*** List et al.: Compared and Contrasted.

| Attribute | List et al. | Present Paper |
|---|---|---|
| Place | University of Central Florida | University of Wyoming |
| *N* (Public) | 44 | 60 and 32 |
| *N* (Private) | 49 | 59 and 37 |
| Subject type | Business School Undergrads | Business School Undergrads |
| Location | Lab | In class |
| Hyp/real | Yes/Yes | No/Yes |
| Elicitation mechanism | Voting referendum | Voting referendum |
| Use randomized response | Yes | Yes |
| Public good | CEPA | WERC |
| Endowed money | Yes (show-up fee) | Yes |
| *p* (vote revelation) | $10/44 \approx 23\%$ | 25% |

*Note: Hyp/real* corresponds to whether or not the studies use hypothetical or consequential treatments. *Randomized response* corresponds to use of a random response technique to generate voting anonymity. *CEPA* is the Center for Environmental Policy Analysis. *WERC* is the Wind Energy Research Center. *Endowed money* indicates whether a subject voted using endowed or earned money. *p(vote revelation)* corresponds to the probability that a subjects vote, in the public treatment, will be made public.

voting behavior of students in a classroom, whereas List et al. make use of a laboratory. While List et al. endow subjects with a \$20.00 show-up fee, we provide subjects with a \$15.00 "participation" fee. We focus our analysis on replicating only the treatments that offer real economic consequences (columns "Actual-RR" and "Actual-PEER" in table 2 of their paper) as our paper is not focused on measuring hypothetical bias but rather on whether people's observed preferences are affected by social pressure, and whether this effect is influenced by group size.

For each treatment, monitors gave subjects \$15.00 of which they decided to keep or donate to the University of Wyoming Wind Energy Research Center (WERC). Subjects voted either "yes" or "no" which corresponded to votes for whether the group should donate or keep its money. The group decision was made by a majority rule. If the referendum passed, the payment was applied to everyone in the group.

In public treatments, the answer sheet included a section in which subjects randomly picked a number between 1 and 100. The 25% of subjects whose number was closest to 1 were chosen to announce the outcome of their vote to the group. This procedure provides risk of vote revelation and is assumed to relieve subjects of social-isolation. For the public treatments, subjects were assured that if they were not randomly chosen to announce their vote, it would remain private from the rest of the group.[3]

In the private treatments, anonymity from the group and from the monitors of the experiment was created using the unrelated question randomized response model of Greenberg, Abul-Ela, Simmons, and Horvitz (1969). This approach has the benefit of creating pure anonymity, however, the technique also increases the standard errors of the estimates. Each participant in the private treatment was given a six-sided die. They were instructed to roll the die one time and fill in the answer sheet appropriately. A roll of a 1 or 2 resulted in a participant answering question A, a roll of 3, 4, 5, or 6 prompted them to answer question B. Question A asked them, "Did you roll a 1?" Question B was the actual valuation question which asked them if they wanted to donate \$15.00 to WERC. To answer either question A or B subjects simply circled either "yes" or "no" at the bottom of the answer sheet. This way, no one, including the monitors of the experiment could identify individual preferences.[4]

Before making the donation decision, monitors gave each subject a description of the WERC:[5]

> Energy in the United States is largely derived from non-renewable resources such as oil, coal and natural gas. Environmental scientists argue that in addition to providing a

> cleaner source of energy, harvesting wind energy is important because of its renewable characteristics; the advancement of wind technology offers a buffer to dwindling stocks of non renewable energy sources. Wind energy has yet to be harvested on mass scales in the United States; the U.S. currently receives less than 1% of its energy from wind. However, the American Wind Energy Association (AWEA) believes by the year 2030, 20% of America's energy could come from wind. According to Mark Northam, director of UW's School of Energy Resources, "important breakthroughs in technology must occur to increase wind energy consumption in the United States". To that regard, the University of Wyoming will soon establish a wind energy research center. The center will provide natural scientists and engineers with the ability to advance wind energy technology. According to the School of Energy Resources (SER) director, the Wind Energy Research Center will be "a program that's probably as impressive as anybody's in the country." The success of the Wind Energy Research Center has and will continue to be dependent on private and public charitable donations. Referencing why the center was not developed years ago, Jonathan Naughton, a University of Wyoming professor of Mechanical Engineering, says that "The problem has always been funding."

Subjects then indicated their donation preference by circling either "yes" or "no" on an answer sheet. Each answer sheet was turned in face down, all votes were counted and subjects were paid.

Following List et al., before each experimental treatment, subjects were assured that donations would be given to WERC. An envelope and a check made out to WERC were presented to the group and it was announced that, "After the experiment has ended, this check will be made out for the appropriate donation amount and anyone can accompany us to mail the check to WERC when the experiment has ended." Following Berrens et al. (1997), subjects were solicited to participate in the experiment during class time and were not paid for participating. The main benefit of this approach is that it minimizes potential selection bias (discussed by, e.g., Harrison, Lau, & Rutström, 2009).

A survey was distributed after the experiment ended that asked subjects to report their age, gender, and approximate parental income. Given the average participant was a 20-year old college student, parental income was decided to be a more revealing statistic. All treatments were run in introductory economics classes in the College of Business. This resulted in subjects being homogeneous in age but diverse in gender and parental income.

## EXPERIMENTAL RESULTS

Table 2 reports the unconditional summary statistics. Figures in parenthesis reflect expected responses. For example, while there were 59 subjects in the private large-group session, we only expect 39 subjects to answer the

***Table 2.*** Unconditional Results.

| Treatment | $N$ | Yes | Proportion Yes (%) |
|---|---|---|---|
| Public, large | 60 | 37 | 61.6 |
| Private, large | 59 | 23 | 38.9 |
| | (39) | (13) | (33.4) |
| Public, small | 32 | 12 | 37.5 |
| Private, small | 37 | 17 | 45.9 |
| | (25) | (11) | (44) |

*Note:* Entries in parenthesis represent total responses net expected random response answers. We expect 1/3 of answers for private treatments to represent answers to the random question (the probability of rolling a 3,4,5 or 6 on a six-sided dice). 1/2 of answers pertaining to the random question are expected to be "yes" and the other half "no" answers.

valuation question due to the use of the anonymity generating, randomized response technique. We arrive at these estimates following Tracy and Fox (1981), in which the probability of a "yes" response is

$$\bar{z} = a\hat{\mu}_x + (1-a)\hat{\mu}_y \tag{5}$$

where $a$ is the probability of answering the valuation question, $\hat{\mu}_x$ is the probability a person answers "yes" to the valuation question, and $\hat{\mu}_y$ is the probability of answering "yes" if a person answers the random, anonymity generating question. Rearranging yields the probability a person answers "yes" to the valuation question

$$\hat{\mu}_x = \frac{\bar{z} - (1-a)\hat{\mu}_y}{a} \tag{6}$$

Because a person answered the valuation question with a 2/3 probability (the probability of rolling a 3, 4, 5, or 6 on a six-sided dice) and a person faced a 1/2 probability of answering "yes" if they answered the random, anonymity generating question (the probability that a person rolled a 1, given they rolled a 1 or 2), the probability a person answers "yes" to the real, valuation question is

$$\hat{\mu}_x = \frac{3}{2}\bar{z} - \frac{1}{4} \tag{7}$$

Our experimental results broadly support the predictions of the model – and support List et al.'s earlier finding. Pooling the data, people in the public treatment were roughly 16% more likely to vote "yes" compared to those in the private treatments.This initial observation is borne out with formal hypothesis testing. Our first null hypothesis is that people in a voting referendum are equally or less likely to vote "yes" in public as they are in private:

$$H_{0,1} : \text{Yes(VRPU)} - \text{Yes(VRPR)} \leq 0$$
$$H_{A,1} : \text{Yes(VRPU)} - \text{Yes(VRPR)} > 0$$

where "VRPU" denotes voting referendum, public elicitation format and "VRPR" denotes voting referendum, private elicitation format. Pooling the data, we reject the null hypothesis that subjects in the public treatment were equally or less likely to vote "yes" as those in the private treatment (significant at the 5% confidence level). Splitting the sample into large- and small-group sessions highlights the effect of group size on social-pressure bias. Subjects in the large-group, referenda sessions ($N \approx 60$) were 28% more likely to vote "yes" in public than in private (significant at the 1% confidence level). This result does not carry over the small-group session though. Rather, the voting behavior of subjects in the small-group sessions ($N \approx 30$) was not affected by the degree of social isolation (see Table 3 for the corresponding $z$-statistics and $p$-values). Alternatively, one may interpret these results as suggesting larger groups produce greater social pressure to vote in favor of a public good. While we acknowledge this potential explanation, seminal work by Asch (1951, 1952), among others, suggests that social impact on individual behavior is constant for group sizes greater than three.

We assess the robustness of the results by conditioning the treatment effect on the observed heterogeneity of the subject pool. We account for

***Table 3.*** Proportions Test.

| Test | $Z$-Statistic | $p$-Value |
|---|---|---|
| VRPU – VRPR ≤ 0, pooled data | 1.74 | .040 |
| VRPU – VRPR ≤ 0, large session | 2.52 | .005 |
| VRPU – VRPR ≤ 0, small session | –.43 | .333 |

*Note:* $Z$-statistics and associated $p$-values pertain to results excluding the expected 1/6 of "yes" and "no" answers from private treatments. We borrow from Tracy and Fox (1981) in our calculation of expected response rates to the real question and corresponding variances.

the randomized response question in private treatments by augmenting the likelihood function according to Berrens et al. (1997). Maximum likelihood estimation was then completed using GAUSS 6.0.

The reference observation is a female with a parental income that exceeds 90,000 U.S. dollar who participated in a public treatment. We again control for the treatment effect by including a dummy variable that takes on the value of 1 if the treatment is private. Pooling the voting referendum treatments, we find that the coefficient on the treatment effect is −.440 (significant at the 10% confidence level). All covariates enter insignificantly. See Table 4 for a detailed set of results.

To further test whether the effect of social isolation is group-size dependent, we split the pooled data into large- and small-group sessions and run individual probit regressions for each. Similar to the earlier results, we find no evidence of social-pressure bias in the small-group session. All other covariates enter insignificantly. The large sample treatment appears to be the driver for the observed social-pressure bias in the pooled referenda data. For the large-group regression, the coefficient on the treatment effect

***Table 4.*** Probit Results.

| Dependent Variable: Percent of Respondents Answering "Yes" | | | |
|---|---|---|---|
| Variable | VR-pooled Coefficient (Std. err.) | VR-small Coefficient (Std. err.) | VR-large Coefficient (Std. err.) |
| Constant | −.172 | −.748 | −1.591 |
| | (.599) | (.893) | (1.229) |
| Treatment | −.440* | .220 | −.990*** |
| | (.241) | (.405) | (.354) |
| Age | .027 | .031 | .125** |
| | (.027) | (.035) | (.063) |
| Male | −.350 | −.246 | −.481 |
| | (.249) | (.396) | (.344) |
| Income | −.068 | −.157 | −.183 |
| | (.231) | (.380) | (.314) |
| Pseudo $R^2$ | .0249 | .037 | .082 |

*Note:* Numbers in parenthesis represent standard errors. *Treatment* is a dummy variable that takes on the value of 1 if the subject participated in a private treatment. *Male* is a dummy variable that takes on the value of 1 if the subject was male. *Income* is a dummy variable that takes on the value of 1 if the parental income of the subject was less than 90,000 U.S. dollar per year. The results from the private treatments are derived using an augmented likelihood function, explained in detail in Berrens et al. (1997).

***, **, * Corresponds to 1%, 5% and 10% significance, respectively.

is −.990 (significant at the 1% confidence level).[6] Taken together, our results broadly support the findings of List et al. – for a sufficiently large-group size, eliciting preferences from subjects in social isolation significantly reduces their observed support for a public good.

## CONCLUSION

Studies have documented a significant "social-pressure" bias. For example, List et al. (2004) find that socially isolated subjects vote "yes" in support of a public good less often by 30%. We develop a structural model that informs the experimental test that follows. A voting referendum may be prone to suffer from social-pressure bias as a voter usually faces a very small probability of being decisive. When faced with a significant chance of vote revelation, a person may vote "yes" to signal support for a public good, knowing good and well her vote doesn't matter in a practical sense. Because the probability of being decisive is decreasing in the size of the voting bloc, social-pressure bias may be more severe when subjects are making decisions within a large group.

Following List et al.'s experimental design as close as possible, we elicit subject preferences for a costly public good in private and public settings, varying the size of the voting bloc from around 30 to 60. Pooling the data, we document a social-pressure bias of 16%. This result is isolated to the large-group treatment. For the small-group session, there is no evidence of a social-pressure bias. For the large-group session ($N \approx 60$), social-pressure bias is remarkably close to that documented by List et al. (about 30%). The theory presented within suggests that this result may be specific to a voting referendum and is not generalizable to other elicitation mechanisms (e.g., dichotomous choice, open-ended questionnaire, etc.). This is a worthy question that we leave to future research.

## NOTES

1. We model utility as a function of audience perception following the existing literature (see, e.g., Della Vigna, List, & Malmendier, 2012; Benabou & Tirole, 2006; Andreoni & Bernheim, 2009).
2. In List et al.'s paper, the private and public treatments are referred to as "randomized response" (RR) and "peer".

3. Social isolation was avoided in List et al.'s public treatment by asking 10 people at random to reveal their vote, which, given their sample size, roughly implied a 25% chance of vote revelation.

4. Given the randomization device, one-third of private treatment subjects answered the random question and which implies one-sixth of all private treatment "yes" and one-sixth of all private treatment "no" answers are random responses.

5. Given the context of the study, we believe it is important that subjects viewed wind energy as a public good. If wind energy technology advancements were viewed as a public bad, it is possible that social pressure actually worked in the opposite direction from that which is expected. Out of sample survey results suggest undergraduate students at the University of Wyoming view wind energy as a public good. One hundred and twenty-one students from introductory economics courses who did not participate in the experiment were surveyed and asked the following question: "Do you consider the advancement of wind energy technology to be a good thing?" Most students (93%) answered "Yes."

6. We recognize that larger sample sizes in referendum treatments increases the probability of rejecting the null hypothesis of minimal social-pressure bias. However, this will always be the case when comparing the results from large and small sample sessions – regardless of the size of the small sample session.

## ACKNOWLEDGMENTS

We thank the Stroock Fund at UW and the Rasmuson Chair at UAA for financial support. Thanks to B. Berrens, J. Kerkvliet, and J. List for their helpful reactions and comments.

## REFERENCES

Andreoni, J., & Bernheim, B. D. (2009). Social image and the 50–50 norm: A theoretical and experimental analysis of audience effects. *Econometrica*, *77*(5), 1607–1636.

Andreoni, J., & Petrie, R. (2004). Public goods experiments without confidentiality: A glimpse into fund-raising. *Journal of Public Economics*, *88*(7–8), 1605–1623.

Asch, S. E. (1951). Effects of group pressure upon the modification and distortion of judgments. In H. Guctzkow (Ed.), *Groups, leadership, and men* (pp. 222–236). Pittsburgh: Carnegie Press.

Asch, S. E. (1952). Group forces in the modification and distortion of judgments. In H. Guctzkow (Ed.), *Social psychology* (pp. 450–473). New York, NY: Prentice Hall.

Beck, N. (1975). A note on the probability of a tied election. *Public Choice*, *23*, 75–79.

Benabou, R., & Tirole, J. (2006). Incentives and pro-social behavior. *American Economic Review*, *96*(5), 1652–1678.

Berrens, R. P., Bohara, A. K., & Kerkvliet, J. (1997). A randomized response approach to dichotomous choice CV. *American Journal of Agricultural Economics*, *79*(1), 252–266.

Bohnet, I., & Frey, B. (1999). The sound of silence in prisoner's dilemma and dictator games. *Journal of Economic Behavior & Organization*, *38*(1), 43–57.

Carson, R. T., & Groves, T. (2007). Incentive and information properties of preference questions. *Environmental and Resource Economics*, *37*(1), 181–210.

Coursey, D., & Schulze, W. (1986). The application of laboratory experimental economics to the contingent valuation of public goods. *Public Choice*, *49*, 47–68.

Cummings, R., Brookshire, D., & Schulze, W. (1986). *Valuing environmental goods: An assessment of the contingent valuation method.* Totowa, NJ: Rowman and Allanheld.

Della Vigna, S., List, J. A., & Malmendier, U. (2012). Testing for altruism and social pressure in charitable giving. *Quarterly Journal of Economics*, 127(1), 1–56.

Fischhoff, B. (2005). Cognitive processes in stated preference methods. In K.-G. Miler & J. Vincent (Eds.), *Handbook of environmental economics* (Vol. 2, pp. 937–968). Amsterdam: Elsevier.

Greenberg, B., Abul-Ela, A., Simmons, W., & Horvitz, D. (1969). The unrelated question: Theoretical framework. *Journal of the American Statistical Association*, *64*(326), 520–539.

Harrison, G. W., Lau, M. I., & Rutström, E. E. (2009). Risk attitudes, randomization to treatment, and self-selection into experiments. *Journal of Economic Behavior and Organization*, *70*(3), 498–507.

List, J., Berrens, R., Bohara, A., & Kerkvliet, J. (2004). Examining the role of social isolation on stated preferences. *American Economic Review*, *94*(3), 741–752.

Rege, M., & Telle, K. (2004). The impact of social approval and framing on cooperation in public good situations. *Journal of Public Economics*, *88*(7–8), 1625–1644.

Shogren, J. (2004). Experimental methods and valuation. In K.-G. Mäler & J. Vincent (Eds.), *Handbook of environmental economics*. Amsterdam: Elsevier.

Shogren, J., & Taylor, L. (2008). On behavioral-environmental economics. *Review of Environmental Economics and Policy*, *2*, 26–44.

Swim, J., Clayton, S., Doherty, T., Gifford, R., Howard, G., Reser, J., … Weber, E. (2009). Psychology an*d global climate change: Addressing a multi-faceted phenomenon and set of challenges.* A report by the american psychological association's task force on the interface between psychology and global climate chang. Retrieved from http://www.apa.org/science/climate-change.

Tracy, P. E., & Fox, J. A. (1981). The validity of randomized response for sensitive measurements. *American Sociological Review*, *60*, 63–69.

# APPENDIX

*Instructions for Public Voting Referendum Treatments*

Energy in the United States is largely derived from nonrenewable resources such as oil, coal and natural gas. Environmental scientists argue that in addition to providing a cleaner source of energy, harvesting wind energy is important because of its renewable characteristics; the advancement of wind technology offers a buffer to dwindling stocks of nonrenewable energy sources. Wind energy has yet to be harvested on mass scales in the United States; the United States currently receives less than 1% of its energy from wind. However, the American Wind Energy Association (AWEA) believes by the year 2030, 20% of America's energy could come from wind. According to Mark Northam, director of UW's School of Energy Resources, "important breakthroughs" in technology must occur to increase wind energy consumption in the United States. To that regard, the University of Wyoming will soon establish a WERC. The center will provide natural scientists and engineers with the ability to advance wind energy technology. According to the SER director, the WERC will be "a program that's probably as impressive as anybody's in the country." The success of the WERC has and will continue to be dependent on private and public charitable donations. Referencing why the center was not developed years ago, Jonathan Naughton, a UW professor of Mechanical Engineering, says that "The problem has always been funding."

This experiment is designed to elicit the value each of you place on the UW WERC. You will be given $15.00 which you must either choose to keep for yourself or donate to the WERC. The decision to keep the money or donate depends on the decision made by the entire group. If more than 50% of you vote to donate the money, each of you must donate your entire $15.00. Conversely, if more than 50% of you choose not to donate the money, no one will donate any money.

The group decision will be made by tallying individual votes of either "yes" or "no." An answer of "yes" is a vote for each person in the group to donate $15.00. An answer of "no" is a vote for each person in the group to keep their $15.00. After the votes have been cast, 25% of you will be chosen at random to reveal your vote, then everyone will return their answer sheets face down. This will assure that only the people running the experiment will know if you answered "yes" or "no." After you return your survey, you will be given a survey that will be used for statistical analysis.

To summarize, if more than 50% of the votes are "no" each of you will be paid $15.00. However, if more than 50% of the participants vote "yes," everyone will donate their $15.00 to the UW WERC and a signed, dated, and addressed check will be filled out for the amount of $15.00 × the number of participants and put in an addressed and stamped envelope. Immediately after the check has been placed in the sealed envelope, anyone of you may accompany me to mail the check to the WERC.

The answer sheet has three boxes on it, a box labeled "yes," a box labeled "no" and box in which you write any number from 1 to 100, which will be used to randomly pick 25% of participants to announce to the group their donation preference.

### *Instructions for Private Voting Referendum Treatments*

Energy in the United States is largely derived from nonrenewable resources such as oil, coal, and natural gas. Environmental scientists argue that in addition to providing a cleaner source of energy, harvesting wind energy is important because of its renewable characteristics; the advancement of wind technology offers a buffer to dwindling stocks of nonrenewable energy sources. Wind energy has yet to be harvested on mass scales in the United States; the United States currently receives less than 1% of its energy from wind. However, the AWEA believes by the year 2030, 20% of America's energy could come from wind. According to Mark Northam, director of UW's School of Energy Resources, "important breakthroughs" in technology must occur to increase wind energy consumption in the United States. To that regard, the University of Wyoming will soon establish a WERC. The center will provide natural scientists and engineers with the ability to advance wind energy technology. According to the SER director, the WERC will be "a program that's probably as impressive as anybody's in the country." The success of the WERC has and will continue to be dependent on private and public charitable donations. Referencing why the center was not developed years ago, Jonathan Naughton, a UW professor of Mechanical Engineering, says that "The problem has always been funding."

This experiment is designed to elicit the value each of you place on the UW wind energy research facility. Each of you will be given $15.00 which you must either choose to keep for yourself or donate to the WERC. The decision to keep the money or donate depends on the decision made by the entire group. If more than 50% of you vote to donate the money, each of you must donate your entire \$15.00. Conversely, if more than 50% of you choose not to donate the money, no one will donate any money.

The group decision will be made by tallying individual votes of either "yes" or "no." A "yes" vote is a vote for each person in the group to donate \$15.00. A "no" vote is a vote for each person in the group to keep their \$15.00. Each of you have been provided with a six-sided die. This die will act as a randomization device which assures voting anonymity. When the experiment begins, each of you will roll the dice once, keeping the value of your roll private. If you roll a 1 or a 2, answer question A, if you role a 3, 4, 5, or 6 answer question B. This assures that the people running the experiment don't know which question you answered and thus, your donation preference.

When the experiment has ended, answer sheets will be turned in face down, assuring anonymity from the group. The people running the experiment will see your answer be we won't know your donation preference because we won't know which question you answered. If more than 50% of the votes are "no" each of you will be paid \$15.00. If more than 50% of the votes are "yes," everyone will donate their \$15.00 to the WERC. The answer sheet has two boxes on it, a box labeled "yes," and a box labeled "no." To answer either question A or B (but not both), simply circle one of the boxes.

After each of you have been paid, please return to your seat to complete a short survey. When the experiment has ended a signed, dated, and addressed check will be filled out for the amount of \$15.00 × the number of participants and put in an addressed and stamped envelope. Immediately after the check has been placed in the sealed envelope, one of you may accompany me to mail the check to the WERC.

Here is an example of how to answer. Suppose you role a 2, this means you must answer question A. Question A asks, "did you roll a 1." You answer "no," and the answer sheet is now completed. If you were to roll a 5, you would answer question "B" which asks you to give your donation preference.

The answer sheet has two boxes on it, a box labeled "yes" and a box labeled "no." To answer either question A or B (but not both), simply circle one of the boxes at the bottom of the page.

# NONVERBAL FEEDBACK, STRATEGIC SIGNALING, AND NONMONETARY SANCTIONING: NEW EXPERIMENTAL EVIDENCE FROM A PUBLIC GOODS GAME

Adam Zylbersztejn


## ABSTRACT

*Recent experiments show that feedback transmission can mitigate opportunistic behavior in repeated social dilemmas. Two nonexcludable explanations have been investigated: strategic signaling and nonmonetary sanctioning. This literature builds on the intuition that under both partner matching (where the same groups of players interact many times) and stranger matching (where groups change continuously), feedback may work as a nonmonetary sanctioning device, but only the former also allows for strategic signaling. Empirical evidence on the two explanations is mixed. Moreover, the usual design may give rise to confounding matching protocol effects.*

*My experiment provides a novel empirical testbed for different channels by which feedback – costless disapproval points – may affect behavior*



**Replication in Experimental Economics**
**Research in Experimental Economics, Volume 18, 153–181**



**ISSN: 0193-2306/doi:10.1108/S0193-230620150000018006**

*in a repeated public goods game. In particular, it is based on a random matching scheme that neutralizes the confounding effects of different matching protocols on behavior.*

*The transmission of feedback is found to foster prosocial behavior. The data favor the nonmonetary sanctioning explanation rather than the signaling hypothesis.*

*This study provides a novel set of evidence that (i) communication may mitigate selfishness in social dilemmas and (ii) the source of this phenomenon may be linked to the emotional reaction that communication evokes in humans.*

**Keywords:** Public goods game; voluntary contribution mechanism; feedback; signaling; nonmonetary sanctioning; disapproval

**JEL classifications:** C72; D83

## INTRODUCTION

In many economic environments, such as online trading, people interact repeatedly, remotely, and anonymously. Nonverbal feedback is often used as an institutional remedy against moral hazard in these situations. For instance, users of online trading platforms (like Amazon or eBay) may rate the other traders in a positive, neutral, or negative way. Moreover, the availability of negative feedback is a key element of such rating systems. For example, Amazon describes the rating system as follows: *we focus on negative feedback ratings as a percentage of total feedback ratings (the "Negative Feedback Rate") as an indicator of seller performance. Our very best sellers have a close to 0% Negative Feedback Rate*. In the same vein, several experimental studies (discussed in the next section) report that (i) institutions based on feedback transmission are capable of enforcing norms of prosocial behavior when individual rationality conflicts with social interest and (ii) negative feedback happens to be a more powerful tool for achieving social efficiency than positive feedback.

Although feedback transmission might be considered as a reputation building mechanism in repeated social dilemmas (Bolton, Katok, & Ockenfels, 2004), recent experiments suggest that it may affect human behavior through different channels: one involves strategic signaling (i.e., how the feedback recipient should act in the future), the other one stems from the idea that agents' utility may not only depend on the monetary gains generated by decisions, but also on the awareness of other people's perception (i.e., approval or disapproval) of these decisions.

This paper further explores the root underlying the behavioral effect of the transmission of nonverbal feedback in repeated social dilemmas and offers a new methodology to disentangle the signaling and the sanctioning function of this mechanism. The experimental design builds on the intuition that under both partner matching (where the same groups of players interact many times) and stranger matching (where groups change after each interaction), feedback may work as a nonmonetary sanctioning device, but only the former also allows for strategic signaling. At the same time, the experimental methodology neutralizes the undesired and potentially confounding effects of the matching protocol on subjects' behavior (Andreoni & Croson, 2008).

In line with some of the previous experiments, the presence of feedback transmitted via costless disapproval points is found to significantly foster prosocial behavior. Furthermore, the data favor the nonmonetary punishment explanation rather than the strategic signaling hypothesis.

## RELATED LITERATURE

In a seminal contribution, Masclet, Noussair, Tucker, and Villeval (2003) use a repeated public goods game, based on the voluntary contribution mechanism (VCM). After each round, every subject observes his group members' contributions, sends a message containing *disapproval points* to each group member, and is informed about the sum of points received from others. Masclet et al. conjecture that the transmission of feedback in a repeated interaction may affect subjects' behavior in two distinct ways. First, it may serve as an information transmission device prior to the next round – for instance, signals may convey a warning that the sender will decrease his future contribution unless the receiver increases his. Second, being aware of others' opinions may operate as a nonmonetary sanction/reward – for instance, people may display an aversion to being disapproved/a preference for being approved, and act less opportunistically so as to avoid/deserve it (Holländer, 1990).[1] They also argue that these two effects may be separated by contrasting subjects' behavior under partner and stranger matching. Intuitively, in the former setting, the effect of such ex post communication may stem from both strategic information transmission and disapproval-aversion, while in the latter behavioral effects may only be a matter of subjects' aversion to disapproval. In their experiment, the presence of communication under partner matching yields significantly higher contributions than under stranger matching, which supports the strategic signaling hypothesis. In a related study, Peeters and Vorsatz (2013) use a similar experimental game and introduce treatments in which every subject may transmit an emoticon (a frowny in one condition, a smiley in the other) to each partner, and then is informed about the number of received emoticons. Comparing patterns of behavior under partner matching (where they observe a moderate, yet statistically insignificant, effect of ex post communication) and stranger matching (where there is virtually no effect of communication), the paper concludes that ex post messages mainly facilitate the exchange of strategic information before an upcoming round.

On the other hand, some experimental studies suggest that the behavioral effect of feedback may be driven by aversion to other's expressed disapproval. López-Pérez and Vorsatz (2010) report that the availability of fixed-form, post-play messages containing judgments of other participants' decisions makes subjects more cooperative in a one-shot prisoner's dilemma game. Ellingsen and Johannesson (2008) and Xiao and Houser (2009) identify the same effect on altruism in subjects playing a dictator game when a free-form, post-play communication is possible. Czap, Czap, Khachaturyan, Burbach, and Lynne (2013) implement a two-stage game in which a common-pool resource is used by a group of subjects, out of which some have private incentives to produce publicly undesirable externalities. They find that the reception of a negative emotional feedback (via frownies) after the first stage reduces externalities in the second stage, while providing positive feedback (via smilies) does the opposite. Dugar (2010) studies a repeated weakest-link coordination game and finds that the introduction of disapproval points helps groups of players converge toward the Pareto-superior Nash equilibrium, while approval points bring no improvement in this regard, inhibiting a fast convergence to the Pareto-inferior Nash equilibrium. Dugar (2013) finds that the availability of disapproval points generates higher contributions than approval points in a fixed-group VCM game based on Masclet et al. (2003), and that combining *both* kinds of points brings a further improvement. Dugar attributes the effects observed in these two experiments to an asymmetrical sensitivity to approval and disapproval.

## EMPIRICAL STRATEGY

The experimental methodology introduced by Masclet et al. (2003) and subsequently used by Peeters and Vorsatz (2013) has its strengths and weaknesses for separating the signaling dimension and the sanctioning dimension of nonverbal feedback in a repeated interaction. Its advantage is that it identifies environments where both of these phenomena either can or cannot co-exist. The main disadvantage is that these environments may affect behavior simultaneously to the content of messages, and that these behavioral effects are highly unpredictable.

As a simple illustration of the above problem, consider a situation where only two factors affect contributions: received feedback and matching protocol. Since both effects arise simultaneously, future contributions are affected not only by past communication but also by the matching protocol. The two effects should not be considered orthogonal: assuming that the content of feedback is correlated with own and others' past behavior (which is the case in most studies of this type of communication, including this one), future messages depend on past contributions, so that the very nature of feedback is also influenced by the matching protocol. Consequently, the matching protocol effect is a confound that aggravates the isolation of the actual relationship between communication and contributions in a repeated game. Thus, the most general solution to this issue is an experimental design in which actions are uncorrelated with the matching scheme.[2]

The central methodological challenge for this study is to create an experimental environment that allows for different features of feedback, at the same time ruling out the unobserved effect of the matching protocol on decisions. To this end, I introduce a novel uniform matching protocol in which groups are maintained or broken up in a random way. In each round of a repeated game, subjects decide upon their level on contribution before learning whether their group prevails until next round. This design neutralizes the unwanted forward-looking effects of different matching protocols on contributions, while controlling for the backward-looking factors (as discussed in the next section). The transmission of feedback, in turn, only takes place after the fate of groups is known to subjects. Consequently, the effect of communication can be captured in two different strategic contexts: when groups prevail from one period to another, and when they change between rounds. In line with Masclet et al.'s original argument, strategic information transmission is only possible in the former case, while nonmonetary punishment may occur in both cases.

Following several previous studies (Dugar, 2010, 2013; Masclet et al., 2003; Noussair & Tucker, 2005), I implement a communication protocol based on numerical ex post messages framed as disapproval points. In contrast to these studies, I use a VCM game that is played by groups of two subjects (instead of four) so as to improve the interpretability of messages.[3] In each round, every participant learns about other group member's contribution, sends a message and receives one in return. Consequently, messages may be easily matched to individual actions, which creates an environment that (i) facilitates agents' comprehension of nonverbal content and (ii) allows the experimenter to trace the relationship between individual messages and individual contributions. Moreover, it is common knowledge that no group ever reappears after having been dissolved, which is aimed at ruling out the possibility of sending future-oriented signals between subjects who are about to cease interacting.

## *Experimental Game and Conditions*

### *Experimental Game*

Pairs of subjects play the following version of the VCM game. Each player holds an initial endowment of 15 Experimental Currency Units (ECU), and may contribute any part of it to the common pool.[4] Decisions are made simultaneously and the amount accumulated in the common pool is then multiplied by 1.5 and retransferred to group members in equal parts. Thus, the gain of player $i$ who contributes $N_i$ and interacts with player $j$ who contributes $M_j$ equals:

$$\text{Gain}_i = 15 - 0.25 \times N_i + 0.75 \times M_j$$

Although the social welfare is maximized when each player contributes his entire endowment, the dominant strategy is to contribute nothing.[5] Therefore, the standard game theory predicts that in the unique Nash equilibrium all players contribute nothing.

### *Baseline Condition*

In the baseline condition (BC), the game is repeated as follows. In each occurrence, subjects (i) decide upon their contribution to the common pool, (ii) learn whether their current groups survive until the next round, and finally (iii) observe the other group member's contribution, as well as their own gain. Subjects are informed that in each round their groups survive with a 50% chance, that this process is entirely random and that every change is permanent – groups that disappear cannot reappear in the future. In all rounds following round 1, an announcement prior to the initial stage reminds subjects whether their group has changed with respect to the previous period. The important issue of the asymmetry of information about players' past behavior between maintained and newly formed groups is addressed in the following way. Before stage (i), members of newly formed pairs are informed about the contribution recently made by their current group member in his former group.[6] Although the extent of historical information may vary between new and old pairs due to the randomness of group rematching, this design maintains the *minimum* level of knowledge about group members' past.[7]

*Communication Condition*

The communication condition (CC) encompasses the three stages forming BC, as well as the current-group-status reminder. In addition, in stage (iv), subjects are asked to express their opinion about a group member's decision by assigning a certain number of points (between 0 and 10) to him. Experimental instructions state that *a high number of points expresses disapproval: 10 points correspond to the strongest disapproval, while 0 points correspond to the weakest disapproval*, and that *assigned points do not affect either participant's gains for the experiment*. Then, each subject is informed about the number of points he received from the other group member. If groups change between periods $t-1$ and $t$, prior to stage (i) subjects not only learn about the decision taken by their current group member in $t-1$ (like in BC), but also about the number of points he received.

## *Experimental Procedures*

The experiment involves a total of 12 sessions (6 for BC and 6 for CC), each comprising 8 subjects. I use a round-robin matching protocol (ensuring that every two subjects have the opportunity to interact during the experiment) and the random group survival rule outlined above. Consequently, the structure of group matching and the number of rounds may differ between sessions. In order to control for the effects of these variations, I use six independent, randomly generated matching sequences (henceforth labeled Game 1, ..., Game 6) and run two separate sessions for each of them: one implementing BC and one implementing CC.[8] The effect of matching sequences is accounted for in the statistical analysis. Otherwise, all sessions are considered independent. Furthermore, all observations (except for those coming from round 1) within each session might be correlated due to the round-robin matching procedure, which is also accounted for in the analysis of data. Subjects are informed that the game contains between 10 and 16 rounds and that its length is determined randomly. In practice, sessions contain between 11 and 15 rounds, and the pairs of subjects interact for up to five consecutive rounds.

At the beginning of each session, participants are randomly assigned to their computers and asked to fill in a small personal questionnaire containing basic questions about their age, gender, education, etc. The written instructions are then read aloud. Before starting, subjects are also asked to fill in a quiz assessing their understanding of the game they are about to play. Once the quiz and all remaining questions are answered, the experiment begins.

Once all pairs complete a round of the game, subjects are informed that either a new round will start or the experiment will end. In the latter case, a single round is randomly drawn and each participant receives the amount in EUR corresponding to his gains in that round (converted from ECU to EUR using an exchange rate 1 ECU = 0.40 EUR), plus a show-up fee equal to 5 EUR.

All sessions took place in the lab of the University of Paris 1 (LEEP) in July 2012. The recruitment of subjects was carried out via LEEP database among individuals who have successfully completed the registration process on laboratory's website.[9] Among 96 participants, 51 are male and 45 are female. Sixty-three participant are students, of which 67% might have some background in game theory due to their field of study.[10] Eighty-two subjects had taken part in experiments organized in LEEP in the past. Participants' average age is roughly 25. No subject participated in more than one experimental session. Each session lasted about 45 minutes with an average earning of €12.20.

## RESULTS

This section establishes the principal results coming from individual- and group-level data analysis. The main findings are as follows. First, the transmission of feedback involving disapproval points fosters the provision of public goods. Second, the effects of ex post communication on group productivity and the within-group structure of contributions do not vary between the cases where messages may or may not have a strategic (i.e., future-oriented) function.

### *Individual Behavior*

Table 1 presents subjects' average contributions as a function of the structure of the experimental game and the CC. In five experimental games out of six, the presence of nonverbal feedback increases the average contribution. This shift in behavior is significant at the 5% level according to a Wilcoxon signed-rank test.[11,12] Then, the behavior of inexperienced players is neutral to the treatment variable: a Wilcoxon–Mann–Whitney rank-sum test using individual observations from round 1 does not reject the hypothesis that the distributions of subjects' choices are the same in both experimental environments ($p = 0.210$).[13]

**Result 1.** The average contribution in the CC is significantly higher than in the BC. However, this effect does not occur in the initial round.

To further investigate the behavioral transition caused by communication, Fig. 1 compares the distributions of contributed amounts in both experimental conditions. In BC, over 40% (251 out of 624) of decisions are

***Table 1.*** Average Contributions by Treatment and Experimental Game.

| Conditions | Game 1 | Game 2 | Game 3 | Game 4 | Game 5 | Game 6 | Overall | *p* |
|---|---|---|---|---|---|---|---|---|
| *Average contributions in round 1* | | | | | | | | |
| Baseline | 7.625 | 6.750 | 8.125 | 6.375 | 5.750 | 3.875 | 6.417 | 0.210 |
| Communication | 6.250 | 8.250 | 6.500 | 8.375 | 7.875 | 8.750 | 7.667 | |
| *Overall average contributions* | | | | | | | | |
| Baseline | 5.000 | 2.900 | 4.420 | 3.083 | 6.010 | 2.846 | 3.942 | 0.046 |
| Communication | 4.083 | 7.783 | 6.045 | 6.325 | 9.135 | 8.029 | 6.938 | |
| *Session details* | | | | | | | | |
| Number of subjects | 8 | 8 | 8 | 8 | 8 | 8 | | |
| Number of rounds | 12 | 15 | 11 | 15 | 12 | 13 | | |

*Note.* Columns 1–6 present average contributions in each experimental game, using data from round 1 (upper part) and all rounds (middle part). The last two columns summarize these results and provide nonparametric tests for the significance of the effect of treatment on contributions: the Wilcoxon–Mann–Whitney rank-sum test using individual observations in round 1, and Wilcoxon signed-rank test using game-level matched averages for the aggregate data (in order to account for the potential matching sequence effects). The lower part of the table contains additional information on the number of subjects and the length of each experimental game.

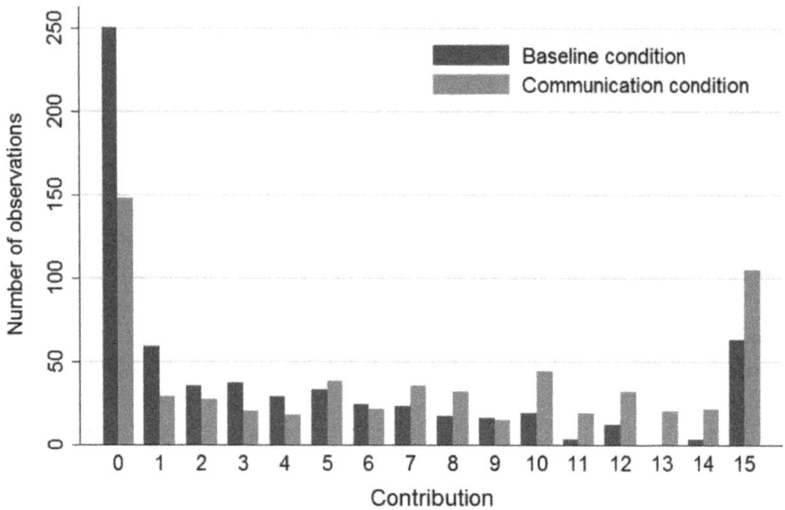


*Fig. 1.* Distribution of Contributions across Treatments. *Note.* For each experimental treatment, data contain 624 observations from 6 experimental sessions. Contributions are given in ECU.

zero contributions, as compared to less than 25% (148 out of 624) in CC. Moreover, contributions between 1 ECU and 4 ECU are more frequent in the former than in the latter. The relationship between the two conditions becomes unstable until the threshold level of 10 ECU, above which all values appear substantially more often in CC than in BC, including the case in which the entire endowment is transferred to the common pool (105 times in CC against 63 times in BC). In addition, Fig. 2 suggests that this phenomenon is robust to the way pairs are matched prior to interacting. Within each experimental condition, one observes similar distributions of contributions in newly formed and retained pairs. On the other hand, the key difference between both conditions observed in Fig. 1 prevails: the frequency of low contributions is substantially lower in CC than in BC, while the opposite holds for high contributions.[14]

### *Group Behavior*

To further uncover the cause of the transition in subjects' decision making, let us now turn to the behavior of groups in which these decisions occur.

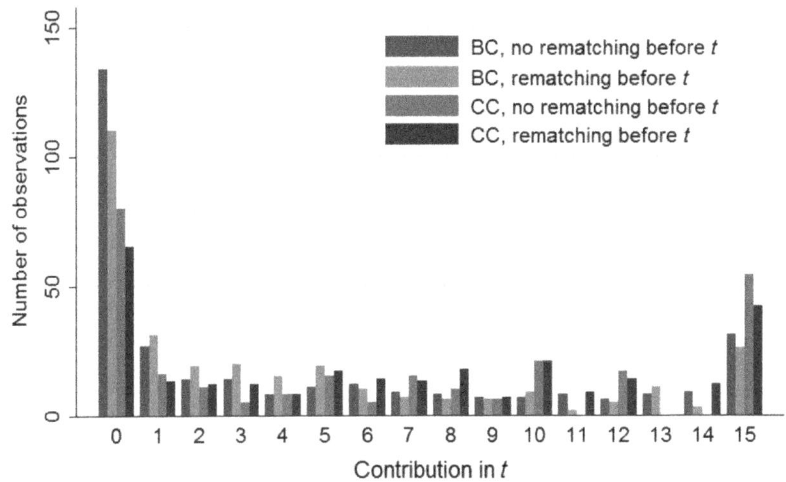


*Fig. 2.* Contributions, Communication, and Rematching. *Note.* For each combination of experimental condition (BC or CC) and pair status (rematching occurs before period t or not), data contain 288 observations from six experimental sessions. Contributions are given in ECU.

This analysis is aimed at providing statistical tests of the signaling and nonmonetary sanctioning hypotheses via parametric regression models. The core instruments are embedded in the 2×2 design of the experiment: the first dimension being the presence of random $1_{\text{Rematching}^{i}_{t-1}}$ (set to 1 if rematching occurs, 0 otherwise) of pairs prior to round $t$, while the second being the presence of ex post communication treatment ($1_{CC} = 1$ for CC, 0 otherwise). The statistical analysis focuses on two dependent variables:

*Contributions gathered by players i and j interacting at time t* ($c_t^i + c_t^j$). The signaling role of ex post communication may be confirmed by the significance of $1_{CC}$ in maintained pairs, and a significant difference between the contributions accumulated in maintained and rematched pairs when $1_{CC} = 1$ – both of which would indicate that the effect of feedback on group productivity depends on the continuity of interaction. On the other hand, a positive effect of ex post communication which does not vary between maintained and rematched pairs would support the nonmonetary sanctioning hypothesis. These mechanisms are tested in Model 1.

*The absolute difference between the contributions of players i and j who form a pair at time t* ($|c_t^i + c_t^j|$). Model 2 provides a complementary test for the signaling hypothesis. If nonverbal feedback operates as an efficient signaling mechanism, then it should facilitate players' mutual strategic adaptation resulting in more equilibrated contributions within pairs. Consequently, the contributions in maintained groups playing under the CC should involve less intra-group inequality than elsewhere.

Table 2 summarizes the estimates from both models.[15] Model 1 reveals that the presence of communication increases the provision of public good in both retained and rematched groups ($H_0$: $\beta_2=0$, $p=0.058$, $H_0$: $\beta_2+\beta_3=0$, $p=0.023$, respectively). Moreover, the group behavior in the CC does not depend on the previous occurrence of rematching ($H_0$: $\beta_1+\beta_3=0$, $p=0.497$). Altogether, this evidence favors the nonmonetary sanctioning explanation over the signaling hypothesis: the impact of communication on the group behavior is significant, but does not depend on players' ability to transmit strategic signals.

Model 2 provides evidence in the same vein: the within-group inequality in maintained pairs does not differ systematically between BC and CC. One cannot reject the joint hypothesis that the level of inequality in maintained pairs is not affected by the presence of ex post communication, and that the level of inequality in the CC does not depend on the nature of groups' rematching ($H_0$: $\beta_2=0 \cap \beta_1+\beta_3=0$, $p=0.464$). On the other hand, the contributions within newly formed groups in the BC tend to be more divergent than in retained groups ($H_0$: $\beta_1=0$, $p=0.099$), which suggests that the experience acquired over the course of a continuous interaction improves players' capacity of strategic adaptation, and that this effect is superseded by the ex post exchange of messages.

**Result 2.** The effect of communication on group productivity and players' strategic adaptation within groups does not depend on the availability of future-oriented signals.

Finally, both regressions also reveal interesting time trends: a gradual decay in the production of public goods in Model 1,[16] and no systematic variations in within-group inequalities in Model 2.

### *Does Rematching Preclude Strategic Signaling? A Simple Test of the Altruistic Signaling Hypothesis*

Note that the underlying assumption of the above analysis is that the transmission of strategic signals is only possible in continuous interactions. However, since players in newly established pairs observe the degree of disapproval their group members received in their old pairs, one may also

***Table 2.*** Determinants of Group Behavior.

| | Model 1 | Model 2 |
|---|---|---|
| Dep. Variable | $c_t^i + c_t^j$ | $\lvert c_t^i - c_t^j \rvert$ |
| | $\beta$ | $\beta$ |
| *Intercept* ($\beta_0$) | 8.032 | 4.422* |
| $1_{\text{Rematching}_{t-1}^i}$ ($\beta_1$) | 0.753 | 0.928* |
| $1_{CC}$ ($\beta_2$) | 6.375* | 0.838 |
| $1_{CC} \times 1_{\text{Rematching}_{t-1}^i}$ ($\beta_3$) | −0.188 | −0.785 |
| Additional controls | | |
| *Round* | | |
| 3 | −1.627 | 1.580 |
| 4 | −0.938 | 0.938 |
| 5 | −2.803** | 0.551 |
| 6 | −2.241** | 0.551 |
| 7 | −2.960** | 0.080 |
| 8 | −3.860** | −0.839 |
| 9 | −3.741*** | 0.592 |
| 10 | −5.205*** | −0.976 |
| 11 | −4.553** | 0.488 |
| 12 | −4.317*** | 0.704 |
| 13 | −4.587** | 0.687 |
| 14 | −3.550** | 1.666 |
| 15 | −4.334** | 0.630 |
| *Game* | | |
| 2 | 2.067 | −0.616 |
| 3 | 1.228 | −1.408 |
| 4 | 0.719 | −1.268 |
| 5 | 6.636 | −0.386 |
| 6 | 2.249 | −1.263 |
| $N$ | 576 | |
| $R^2$ | 0.208 | 0.043 |

*Note.* OLS regressions using group-level observations. Dependent variable in Model 1 is the sum of contributions of players $i$ and $j$ forming a group at time $t$ ($c_t^i + c_t^j$), whereas Model 2 uses the absolute difference of contributions within a group ($\lvert c_t^i - c_t^j \rvert$). Columns contain model's coefficients ($\beta$) and corresponding $p$-values, with */**/*** indicating significance at the 10%/5%/1% level. Explanatory variables include a dummy indicating the occurrence of group rematching prior to ($1_{\text{Rematching}_{t-1}^i}$) and another dummy indicating whether the game involves ex post communication ($1_{CC}$), as well as their interaction. Models also control for the round and matching-sequence effects, with *Round 2* and *Game 1* being set as reference points and omitted in regressions. Data include observation from round 2 onwards. Standard errors are clustered at the session level (12 clusters) and are corrected using leave-one-out jackknife.

consider an alternative scenario – altruistic signaling – where disapproval points are utilized to label players who are about to enter into a new interaction. More precisely, subsequent members of each subject's pair may transmit cues about this person to their successors and adapt their decisions according to the cues transmitted by their predecessors.

To test this hypothesis, the data from newly matched pairs are used to regress the individual contributions on the historical information held by each player about his current group member: his past contribution and the number of disapproval points he received in the previous period. The estimates are presented in Table 3. This simple empirical test suggests that rematching precludes the strategic use of these points: while group members' past behavior matters for subjects' choices, the number of disapproval points the former received in their previous groups does not.

### *Determinants of Expressed Disapproval*

This section addresses two central questions related to the formation of ex post messages: (i) what is the relationship between the content of feedback and underlying contributions and (ii) does this relationship vary between retained and rematched groups? The answer to the first question is provided by Fig. 3, which presents the average number of points sent by a subject to his group member as a function of his deviation from the latter's contribution, and the outcome of the random draw that determines the future matching. Under both matching scenarios, the overall pattern seems

***Table 3.*** Contributions and Historical Information in Newly Formed Pairs.

| | | | |
|---|---|---|---|
| *Intercept* ($\theta_0$) | 4.133*** | 7.200*** | 4.304** |
| *Group_member's_contribution*$_{t-1}$ ($\theta_1$) | 0.228*** | – | 0.291** |
| *Group_member's_received_points*$_{t-1}$ ($\theta_2$) | – | −0.080 | 0.158 |
| *N* | | 288 | |
| $R^2$ | 0.053 | 0.003 | 0.060 |

*Note*. OLS regressions of player's contribution in $t$ on the historical information about the current group member: his contribution in $t-1$ and the number of points he received in $t-1$. Columns contain model's coefficients ($\theta$), with */**/*** indicating significance at the 10%/5%/1% level. Data come from newly matched pairs. Standard errors are estimated using observations clustered at the session level (six clusters in total) and corrected using leave-one-out jackknife.

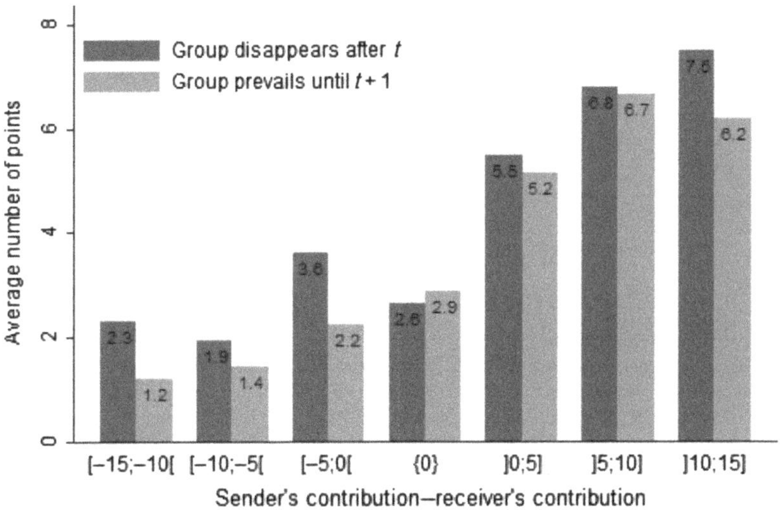


*Fig. 3.* Sent Disapproval Points and the Relative Size of Contribution. *Note.* Data come from six sessions and contain 288 observations for each type of group. Contributions are given in ECU.

rather coherent: the number of sent disapproval points grows as the difference between sender's and receiver's contributions increases.

Answering the second question requires a more formal insight into the process of message formation. For this sake, I estimate a linear model representing the number of points sent by a player in round $t$ as a function of his $Group\_member's\_contribution_t$, as well as its relative size within his group – that is, the magnitude of $Sender's\_positive\_deviation_t$ or $Sender's\_absolute\_negative\_deviation_t$ with respect to other group member's decision. The effect of prospective rematching is captured by the variable $1_{Rematching_t^{t+1}}$ (like before, set to 1 if rematching is about to occur between $t$ and $t+1$, a 0 otherwise) as well as its interactions with the above variables. The estimates are presented in Table 4.

In the absence of rematching prior to the upcoming interaction, players send more disapproval points the less their group members contributed ($H_0$: $\gamma_1 = 0$, $p = 0.002$), and the more they contributed themselves relative to their group members ($H_0$: $\gamma_2 = 0$, $p = 0.010$). The magnitude of own *negative* deviation (in absolute terms), in turn, does not play a significant role in the process of point attribution ($H_0$: $\gamma_3 = 0$, $p = 0.329$). Importantly, these

**Table 4.** Sent Disapproval Points and the Relative Size of Contribution.

| | |
|---|---|
| *Intercept* ($\gamma_0$) | 6.227*** |
| *Group_member's_contribution*$_t$ ($\gamma_1$) | −0.380*** |
| *Sender's_positive_deviation*$_t$ ($\gamma_2$) | 0.167*** |
| *Sender's_absolute_negative_deviation*$_t$ ($\gamma_3$) | 0.075 |
| $1_{\text{Rematching}_t^{t+1}}$ ($\gamma_4$) | −0.536 |
| $1_{\text{Rematching}_t^{t+1}}$ × *Group_member's_contribution*$_t$ ($\gamma_5$) | 0.047 |
| $1_{\text{Rematching}_t^{t+1}}$ × *Sender's_positive_deviation*$_t$ ($\gamma_6$) | 0.010 |
| $1_{\text{Rematching}_t^{t+1}}$ × *Sender's_absolute_negative_deviation*$_t$ ($\gamma_7$) | −0.138 |
| $N$ | 576 |
| $R^2$ | 0.361 |

*Note.* OLS regressions of the number of points sent in $t$ on variables indicating the level of *Group_member's_contribution*$_t$, the value of Sender's_positive_deviation$_t$ or *Sender's_absolute_negative_deviation*$_t$, a dummy indicating the occurrence of $1_{\text{Rematching}_t^{t+1}}$ as well as its interactions with the three above variables. */**/*** indicate coefficients' significance at the 10%/5%/1% level. Standard errors are estimated using observations clustered at the session level (six clusters in total) and corrected using leave-one-out jackknife.

effects do not change due to rematching. One may not reject the joint hypothesis that the prospect of group rematching does not change the relationship between one's relative contribution and the number of disapproval points he sends ($H_0$: $\gamma_5 = 0 \cap \gamma_6 = 0 \cap \gamma_7 = 0$, $p = 0.356$).

**Result 3.** Ex post disapproval points are assigned in an informative and coherent manner: low contributors receive more disapproval points than high contributors. This pattern of point attribution does not vary significantly between pairs that are about to cease and continue interacting.

## SUMMARY AND DISCUSSION

Recent developments in economic research suggest that communication may mitigate selfishness in social dilemmas, and that the source of this phenomenon lies in the emotional reaction that communication evokes in humans. For instance, Ellingsen and Johannesson (2004) and Vanberg (2008) argue that ex ante communication may reduce the scope of opportunistic behavior due to agents' *aversion to lying*, while Charness and Dufwenberg (2006) relate this transition to *guilt aversion* due to which agents experience disutility from letting down others' expectations.

The present study provides new evidence on the behavioral effect of communication in social dilemmas. First, echoing the previous findings by Masclet et al. (2003) and Dugar (2013) (and notwithstanding the findings by Peeters & Vorsatz, 2013), the paper reports that the availability of a feedback mechanism involving costless disapproval points substantially reduces opportunistic behavior in a repeated VCM game. Second, it offers a new angle for understanding this behavioral process by disentangling its potential (and nonexcludable) driving forces: first, the aversion to disapproval, understood as nonmonetary disutility drawn from others' opinions about own behavior; second, the transmission of strategic signals (such as threats) linked to future interactions. It builds on the intuition that under both a repeated interaction within fixed groups of players (partner protocol) and interactions within constantly changing groups (stranger protocol) ex post communication may be used for nonmonetary sanctioning or rewarding, but only the former also allows for future-oriented strategic signaling. While retaining these different features of communication, this new design neutralizes the undesired effects of different matching protocols on contributions.

The experimental literature lacks unequivocal evidence on the root underlying the behavioral effect of ex post communication in repeated social dilemmas. For instance, Masclet et al. (2003) suggest that both of these mechanisms may influence subjects' behavior, Peeters and Vorsatz (2013) highlight the role of strategic information transmission, while Dugar (2013) reports that one's utility might be sensitive to other people's judgement of one's actions. The data from this experiment do not support the signaling hypothesis, favoring the nonmonetary punishment explanation instead. First, the provision of public good is invariant across groups that may exchange future-oriented messages and those whose messages cannot be future oriented. Ex post communication increases the production of public good irrespectively of the continuity of group's interaction. However, it fails to improve players' strategic adaptation (measured as the within-group inequality of contributions) even if interactions extend throughout multiple periods. Second, the relationship between one's relative contribution and the resulting level of others' disapproval is coherent (low contributors face stronger disapproval than high contributors) and does not vary as a function of the continuity of interaction.

Finally, this experiment contributes to the recent investigations of verbal and nonverbal feedback mechanisms in social dilemmas. On the one hand, several studies suggest that ex post communication based on natural language affects subjects' emotions in social dilemmas (Ellingsen &

Johannesson, 2008; López-Pérez & Vorsatz, 2010; Xiao & Houser, 2009). On the other hand, Dugar (2010, 2013) suggests that nonverbal ex post communication may also refer to emotions. The results of this experiment are in line with the latter finding. Adding a structured and wordless procedure of expressing judgments creates an environment in which aversion to others' disapproval may refrain opportunistic behavior in social dilemmas. A more systematic comparison of verbal and nonverbal feedback mechanisms could certainly constitute an avenue for future research.

# NOTES

1. Masclet et al. (2003) refer to ex post communication as *nonmonetary sanctioning*, since they also investigate another treatment where points are related to monetary punishment. A similar nomenclature (like *informal sanctions* or *nonmonetary punishment*) is also adopted in the other papers discussed in this section. Since the goal here is to test whether this institution is actually used to inform or to sanction people, I consistently use a more neutral term – ex post communication.

2. Masclet et al. (2003) offer a way to control for the matching protocol effects with a difference-in-difference approach: in every session, subjects play a sequence of rounds without communication, which is followed by an analogous sequence with communication, and then another sequence without communication, holding the matching protocol constant. They argue that since subjects' behavior is similar in the initial sequence under both matching schemes, thus the differences observed in the second stage are unlikely to stem from the matching protocol effects. However, it should also be noticed that a conclusive comparison in the second stage is only possible in the absence of a systematic partner–stranger difference in the first stage. Unfortunately, scenarios where observations are neutral to matching protocol effects are far from being a regularity in lab experiments on public goods games. In a related study, Peeters and Vorsatz (2013) use a classical between-subject design and observe systematic matching protocol effects: in each round of every treatment, partner matching induces higher contributions than stranger matching (see the working paper version of their study, Peeters & Vorsatz, 2009). Moreover, the matching protocol effects are absent in only 4 out of 13 experiments overviewed by Andreoni and Croson (2008). In five cases, partner matching brings higher contribution than stranger matching, whereas the opposite occurs in the remaining four cases.

3. Although Isaac and Walker (1988) and Isaac, Walker, and Williams (1994) suggest that the group size does not affect contributions in VCM, one might nonetheless argue that it could change the role of communication in favor of strategic signaling. First, the improved clarity of communication between players might facilitate the exchange of strategic signals. Second, the pressure induced by nonmonetary sanctions may be weaker in smaller groups. However, such scenario is not supported by the data which reveal a predominant role of nonmonetary sanctioning in the two-person game.

4. In the lab implementation, contributions may only take integer values between 0 ECU and 15 ECU.

5. To avoid framing effects, instructions use neutral phrasing: I use expressions such as *players* and *group members* rather than *partners*, and *contributions* are never related to *cooperation*. See Rege and Telle (2004) for evidence on framing effects in public goods experiments.

6. Instructions (translated from French to English) are provided as a supplementary material in Appendix A.

7. One may argue that this design may induce reputation-building. However, even if subjects indeed try to establish a reputation, this incentive remains constant throughout the entire experiment: across different kinds of pairs – newly established and preserved, as well as under different communication conditions. Yet, this symmetry is another important refinement with respect to the usual partner–stranger comparison in which reputation-building may unobservedly arise in the former scheme, but not in the latter.

8. The exact structure of these sequences can be found in Appendix B.

9. The recruitment uses Greiner (2004); the experiment is computerized through a software developed under Regate (Zeiliger, 2000).

10. Disciplines such as economics, engineering, management, political science, psychology, mathematics applied in social science, mathematics, computer science, sociology, and biology.

11. For each of the six games, this test matches the average contributions observed in BC and in CC, and therefore accounts for the effects related to different matching sequences.

12. All $p$-values used in statistical analysis correspond to two-sided tests.

13. This may suggest that the effect of communication via disapproval points involves learning of conventions and arises via procedural experience. An identical phenomenon is reported by Dugar (2010, 2013).

14. For instance, zero contributions are observed 134 times in BC in the absence of prior rematching of pairs, and 110 times had rematching occurred. Analogous figures in CC are 80 and 65. Moving to the opposite extreme, contributions of 15 occur 31, 26, 54, and 42 times in the four respective cases.

15. Appendix C provides an additional specification – double-censored tobit – as an alternative to the OLS estimates reported in Tables 2–4. Both methods yield highly consistent results.

16. This observation echoes previous experiments on public goods with and without feedback mechanisms. See, for instance, Masclet et al. (2003); Dugar (2013); Peeters and Vorsatz (2013).


## ACKNOWLEDGMENTS

This paper previously circulated under the title "Strategic signaling or emotional sanctioning? An experimental study of ex post communication in a repeated public goods game" and is a thoroughly revised version of CES

WP 2013.11 and WU Wien Economics WP 161. I wish to thank Enrique Fatas and an anonymous referee for helpful suggestions, and Juergen Bracht, Tamás Csermely, Guillaume Fréchette, Yukihito Funaki, Daniel Houser, Nicolas Jacquemet, Frédéric Koessler, Martin Leroch, David Masclet, Ernesto Reuben, Rupert Sausgruber, Thomas A. Stephens, Jean-Marc Tallon, Antoine Terracol, Christian Traxler, Marie Claire Villeval, and Marc Willinger for inspiring discussions at various stages of my work. Participants at 2013 Mainz Workshop "Microfoundations of group conflict," 2013 International Meeting of the Association for Public Economic Theory in Lisbon and 2014 ADRES PhD Meeting in Paris delivered helpful comments. Maxim Frolov and Ivan Ouss provided assistance in running the experiments. This work was supported by the French National Research Agency (ANR), through the program Investissements d'Avenir (ANR-10−LABX_93-01). This research was performed within the framework of the LABEX CORTEX (ANR-11-LABX-0042) of Université de Lyon, within the program Investissements d'Avenir (ANR-11-IDEX-007) operated by the French National Research Agency (ANR).

## REFERENCES

Andreoni, J., & Croson, R. (2008). Partners versus strangers: Random rematching in public goods experiments. In C. R. Plott & V. L. Smith (Eds.), *Handbook of experimental economics results* (Vol. 1, chapter 82, pp. 776−783). North-Holland: Elsevier.

Bolton, G. E., Katok, E., & Ockenfels, A. (2004). How effective are electronic reputation mechanisms? An experimental investigation. *Management Science*, *50*(11), 1587−1602.

Charness, G., & Dufwenberg, M. (2006). Promises and partnership. *Econometrica*, *74*(6), 1579−1601.

Czap, H. J., Czap, N. V., Khachaturyan, M., Burbach, M. E., & Lynne, G. D. (2013). Smiley or Frowney: The effect of emotions and framing in a downstream water pollution game. *International Journal of Economics and Finance*, *5*(3), 9−23.

Dugar, S. (2010). Nonmonetary sanctions and rewards in an experimental coordination game. *Journal of Economic Behavior & Organization*, *73*(3), 377−386.

Dugar, S. (2013). Non-monetary incentives and opportunistic behavior: Evidence from a laboratory public good game. *Economic Inquiry*, *51*(2), 1374−1388.

Ellingsen, T., & Johannesson, M. (2004). Promises, threats and fairness. *Economic Journal*, *114*(495), 397−420.

Ellingsen, T., & Johannesson, M. (2008). Anticipated verbal feedback induces altruistic behavior. *Evolution and Human Behavior*, *29*(2), 100−105.

Greiner, B. (2004). An online recruitment system for economic experiments. In K. Kremer & V. Macho (Eds.), *Forschung und wissenschaftliches Rechnen 2003* (Vol. 63, pp. 79−93). GWDG Bericht. Göttingen: Datenverarbeitung. Ges. für Wiss.

Holländer, H. (1990). A social exchange approach to voluntary cooperation. *American Economic Review*, *80*(5), 1157–1167.

Isaac, R. M., & Walker, J. M. (1988). Group size effects in public goods provision: The voluntary contributions mechanism. *The Quarterly Journal of Economics*, *103*(1), 179–199.

Isaac, R. M., Walker, J. M., & Williams, A. W. (1994). Group size and the voluntary provision of public goods: Experimental evidence utilizing large groups. *Journal of Public Economics*, *54*(1), 1–36.

López-Pérez, R., & Vorsatz, M. (2010). On approval and disapproval: Theory and experiments. *Journal of Economic Psychology*, *31*(4), 527–541.

Masclet, D., Noussair, C., Tucker, S., & Villeval, M. C. (2003). Monetary and nonmonetary punishment in the voluntary contributions mechanism. *American Economic Review*, *93*(1), 366–380.

Noussair, C., & Tucker, S. (2005). Combining monetary and social sanctions to promote cooperation. *Economic Inquiry*, *43*(3), 649–660.

Peeters, R., & Vorsatz, M. (2009). Immaterial rewards and sanctions in a voluntary contribution experiment. METEOR DP 005.

Peeters, R., & Vorsatz, M. (2013). Immaterial rewards and sanctions in a voluntary contribution experiment. *Economic Inquiry*, *51*(2), 1442–1456.

Rege, M., & Telle, K. (2004). The impact of social approval and framing on cooperation in public good situations. *Journal of Public Economics*, *88*(7–8), 1625–1644.

Vanberg, C. (2008). Why do people keep their promises? An experimental test of two explanations. *Econometrica*, *76*(6), 1467–1480.

Xiao, E., & Houser, D. (2009). Avoiding the sharp tongue: Anticipated written messages promote fair economic exchange. *Journal of Economic Psychology*, *30*(3), 393–404.

Zeiliger, R. (2000). *A presentation of regate, internet based software for experimental economics*. Retrieved from http://www.gate.cnrs.fr/zeiliger/regate/RegateIntro.ppt. GATE.

# APPENDIX A

*Written Instructions*

*Author's note: BC (CC) at the beginning of a paragraph indicates that this part is specific to baseline condition (communication condition)*

You are about to take part in an experiment in which you can earn money. Your gains will depend on your decisions, as well as on the decisions made by other participants.

Before starting, we would like to ask you to answer a few standard questions (concerning your age, education, profession, etc.) which will help us to get to know you better. **This information, as well as the amount of your gains in this experiment, will remain strictly confidential and anonymous.**

Please, fill in the questionnaire using the interface on your computer screen, which is divided into three parts:

- In the *top* section, you will find information that might help you in making decisions.
- In the *middle* section, you will submit your decisions by clicking on a relevant button.
- In the *bottom* section, you will see all your decisions and gains from previous rounds of the experiment.

**Thank You.**

*The Experiment*

The experiment consists of several rounds. The total number of rounds is random and might vary between 10 and 16. In each round, participants are divided into groups of two. In each round (more precisely, in Stage 2 described below), a random draw decides that:

- either your group will not change in the next round;
- or that your group will change in the next round. In order to form your new group, another participant will be chosen at random among participant who have never been part of your group before.

**Both Events Are Equally Probable – Each Occurs with a 50% Chance**

In every round, each participant's gain is determined as follows. At the beginning of round, every person possesses the initial endowment of 15 ECU.

Members of each group may create a common pool. Each participant freely chooses his level of contribution to the pool that may range between 0 ECU and 15 ECU. The total amount gathered in the pool is then multiplied by 1.5 and divided equally among group members.

For instance, if you are participant *i* who contributed $N_i$, and the other member of your group, participant *j*, contributed $M_j$, then your gain ($\text{Gain}_i$) equals:

$$\text{Gain}_i = 15 - 0.25 \times N_i + 0.75 \times M_j$$

Table A1 presents your gain in ECU in a given period as a function of your level of contribution and the other group member's level of contribution.

***Table A1.*** Your Gain in ECU in a Given Period As a Function of Your Decision and Your Group Member's Decision.

| Your Level of Contribution | Your Group Member's Level of Contribution | | | | | | | | | | | | | | | |
|---|---|---|---|---|---|---|---|---|---|---|---|---|---|---|---|---|
| | 0 | 1 | 2 | 3 | 4 | 5 | 6 | 7 | 8 | 9 | 10 | 11 | 12 | 13 | 14 | 15 |
| 0 | 15 | 15.75 | 16.5 | 17.25 | 18 | 18.75 | 19.5 | 20.25 | 21 | 21.75 | 22.5 | 23.25 | 24 | 24.75 | 25.5 | 26.25 |
| 1 | 14.75 | 15.5 | 16.25 | 17 | 17.75 | 18.5 | 19.25 | 20 | 20.75 | 21.5 | 22.25 | 23 | 23.75 | 24.5 | 25.25 | 26 |
| 2 | 14.5 | 15.25 | 16 | 16.75 | 17.5 | 18.25 | 19 | 19.75 | 20.5 | 21.25 | 22 | 22.75 | 23.5 | 24.25 | 25 | 25.75 |
| 3 | 14.25 | 15 | 15.75 | 16.5 | 17.25 | 18 | 18.75 | 19.5 | 20.25 | 21 | 21.75 | 22.5 | 23.25 | 24 | 24.75 | 25.5 |
| 4 | 14 | 14.75 | 15.5 | 16.25 | 17 | 17.75 | 18.5 | 19.25 | 20 | 20.75 | 21.5 | 22.25 | 23 | 23.75 | 24.5 | 25.25 |
| 5 | 13.75 | 14.5 | 15.25 | 16 | 16.75 | 17.5 | 18.25 | 19 | 19.75 | 20.5 | 21.25 | 22 | 22.75 | 23.5 | 24.25 | 25 |
| 6 | 13.5 | 14.25 | 15 | 15.75 | 16.5 | 17.25 | 18 | 18.75 | 19.5 | 20.25 | 21 | 21.75 | 22.5 | 23.25 | 24 | 24.75 |
| 7 | 13.25 | 14 | 14.75 | 15.5 | 16.25 | 17 | 17.75 | 18.5 | 19.25 | 20 | 20.75 | 21.5 | 22.25 | 23 | 23.75 | 24.5 |
| 8 | 13 | 13.75 | 14.5 | 15.25 | 16 | 16.75 | 17.5 | 18.25 | 19 | 19.75 | 20.5 | 21.25 | 22 | 22.75 | 23.5 | 24.25 |
| 9 | 12.75 | 13.5 | 14.25 | 15 | 15.75 | 16.5 | 17.25 | 18 | 18.75 | 19.5 | 20.25 | 21 | 21.75 | 22.5 | 23.25 | 24 |
| 10 | 12.5 | 13.25 | 14 | 14.75 | 15.5 | 16.25 | 17 | 17.75 | 18.5 | 19.25 | 20 | 20.75 | 21.5 | 22.25 | 23 | 23.75 |
| 11 | 12.25 | 13 | 13.75 | 14.5 | 15.25 | 16 | 16.75 | 17.5 | 18.25 | 19 | 19.75 | 20.5 | 21.25 | 22 | 22.75 | 23.5 |
| 12 | 12 | 12.75 | 13.5 | 14.25 | 15 | 15.75 | 16.5 | 17.25 | 18 | 18.75 | 19.5 | 20.25 | 21 | 21.75 | 22.5 | 23.25 |
| 13 | 11.75 | 12.5 | 13.25 | 14 | 14.75 | 15.5 | 16.25 | 17 | 17.75 | 18.5 | 19.25 | 20 | 20.75 | 21.5 | 22.25 | 23 |
| 14 | 11.5 | 12.25 | 13 | 13.75 | 14.5 | 15.25 | 16 | 16.75 | 17.5 | 18.25 | 19 | 19.75 | 20.5 | 21.25 | 22 | 22.75 |
| 15 | 11.25 | 12 | 12.75 | 13.5 | 14.25 | 15 | 15.75 | 16.5 | 17.25 | 18 | 18.75 | 19.5 | 20.25 | 21 | 21.75 | 22.5 |

### *What Happens in Each Round*

At the beginning of each round (accept for the first one), the result of the random draw that took place in **the previous period** is recalled to each participant. Each participant is informed that:

- either his group remains the same as in the previous round;
- *BC:* or in the ongoing round, he will play with a different person. In this case, each participant is also informed about the level of contribution of his current group member in the previous period.
- *CC:* or in the ongoing round, he will play with a different person. In this case, each participant is also informed about the level of contribution of his current group member and the number of points he received in the previous period.

Every round contains three stages:

*Stage 1.* Each participant chooses his level of contribution to the pool.
*Stage 2.* Each participant observes the outcomes of a random draw that determines his group in the next round.
*Stage 3. BC:* Finally, every participant is informed about his group member's level of contribution and his own gain for the round.
*CC*:

- At the beginning of this stage, every participant is informed about his group member's level of contribution and his own gain for the round.
- Then, each participant has possibility to express his opinion about the other group member by assigning him a certain number of points. **A high number of points expresses disapproval: 10 points correspond to the strongest disapproval, 0 points correspond to the weakest disapproval**. To do this, you should select the number of points on your screen and press OK button to confirm your choice. The choice of the number of points **has no impact on either participants' gains in the experiment**.
- Finally, **every participant is informed about the number of points that was assigned to him** by the other group member.

At the end of each round, a message on your screen will inform you that either a new round is about to start, or that the experiment ends.

*Payment of Your Earnings*

At the end of the experiment, **one round is selected at random**. Each participant receives a sum in EUR corresponding to the amount he earned in this round, converted from ECU to EUR using an exchange rate 1 ECU = 0.40 EUR, plus a bonus of €5 for completing the experiment. All payments are made individually and in cash.

For obvious reasons, **you are not allowed to talk during the experiment**. Participants who violate this rule will be excluded from the experiment and all payments. It is crucial that you understand perfectly the rules of this experiment. Should you have any questions to ask, please raise your hand.

**Thank you for your participation!**

# APPENDIX B

## *Evolution of Contributions in Experimental Games*

Fig. B1(a)–(f) suggests that treatment effect is stable over time: in Games 2, 5, and 6 contributions in CC dominate those in BC in all rounds; in Games 3 and 4 this tendency is reversed in only 3 rounds out of 11, and 1 round out of 15, respectively. Only in Game 1, the relationship is more ambiguous.

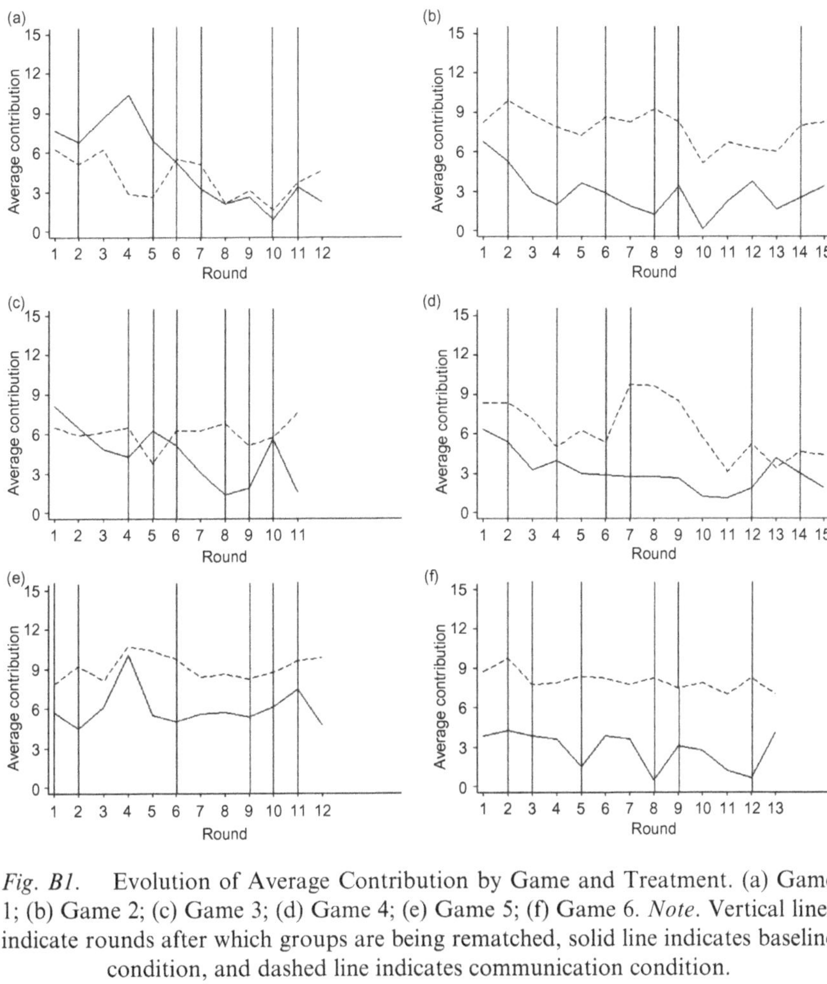


*Fig. B1.* Evolution of Average Contribution by Game and Treatment. (a) Game 1; (b) Game 2; (c) Game 3; (d) Game 4; (e) Game 5; (f) Game 6. *Note*. Vertical lines indicate rounds after which groups are being rematched, solid line indicates baseline condition, and dashed line indicates communication condition.

# APPENDIX C

*Robustness Check: OLS versus Double-Censored Tobit Regressions*

This section provides a comparison of different specifications (OLS vs. doubled-censored tobit) of regression models presented in Tables 2–4. For each model, the columns contain the coefficients from an OLS regression, as well as the coefficients and average marginal effects (ME) from a double-censored tobit regression. */**/*** indicate statistical significance at the 10%/5%/1% level. Altogether, this evidence suggests that both estimation methods yield similar results, with OLS appearing as more conservative in the assessment of the treatment effects and therefore used in this paper.

***Table C1.*** Determinants of Group Behavior – Extension of Table 2.

| | Model 1 | | | Model 2 | | |
|---|---|---|---|---|---|---|
| Dep. Variable | $c_t^i + c_t^j$ | | | $\lvert c_t^i - c_t^j \rvert$ | | |
| | $\beta_{OLS}$ | $\beta_{Tobit}$ | $ME_{Tobit}$ | $\beta_{OLS}$ | $\beta_{Tobit}$ | $ME_{Tobit}$ |
| *Intercept* ($\beta_0$) | 8.032 | 6.741 | – | 4.422* | 3.534 | – |
| $1_{Rematching_{t-1}^i}$ ($\beta_1$) | 0.753 | 1.996* | 1.365* | 0.928* | 1.659* | 1.184** |
| $1_{CC}$ ($\beta_2$) | 6.375* | 7.652* | 6.177** | 0.838 | 1.229 | 0.877 |
| $1_{CC} \times 1_{Rematching_{t-1}^i}$ ($\beta_3$) | −0.188 | −1.013 | −0.816 | −0.785 | −1.311 | −0.935 |
| Additional controls | | | | | | |
| *Round* | | | | | | |
| 3 | −1.627 | −2.211* | −1.873** | 1.580 | 1.979 | 1.467 |
| 4 | −0.938 | −1.208 | −1.029 | 0.938 | 0.911 | 0.659 |
| 5 | −2.803** | −3.171** | −2.670** | 0.551 | 0.537 | 0.385 |
| 6 | −2.241** | −2.740** | −2.313* | 0.551 | 0.708 | 0.510 |
| 7 | −2.960** | −3.517** | −2.952*** | 0.080 | 0.061 | 0.043 |
| 8 | −3.860** | −4.883** | −4.050*** | −0.839 | −1.359 | −0.916 |
| 9 | −3.741*** | −4.788*** | −3.975*** | 0.592 | 0.469 | 0.335 |
| 10 | −5.205*** | −6.339*** | −5.175*** | −0.976 | −1.614 | −1.077 |
| 11 | −4.553** | −5.694** | −4.683*** | 0.488 | 0.434 | 0.310 |
| 12 | −4.317*** | −5.594*** | −4.606*** | 0.704 | 0.602 | 0.433 |
| 13 | −4.587** | −5.693** | −4.683*** | 0.687 | 0.508 | 0.364 |
| 14 | −3.550** | −5.123** | −4.241*** | 1.666 | 1.326 | 0.969 |
| 15 | −4.334** | −5.829** | −4.787*** | 0.630 | 0.501 | 0.359 |
| *Game* | | | | | | |
| 2 | 2.067 | 2.569 | 2.019 | −0.616 | −0.688 | −0.504 |
| 3 | 1.228 | 1.624 | 1.258 | −1.408 | −1.603 | −1.146 |
| 4 | 0.719 | 1.421 | 1.098 | −1.268 | −1.139 | −0.825 |
| 5 | 6.636 | 8.013 | 6.633 | −0.386 | −0.328 | −0.242 |
| 6 | 2.249 | 2.789 | 2.198 | −1.263 | −1.265 | −0.913 |
| *N* | | | 576 | | | |

***Table C2.*** Contributions and Historical Information in Newly Formed Pairs – Extension of Table 3.

| | $\theta_{OLS}$ | $\theta_{Tobit}$ | $ME_{Tobit}$ | $\theta_{OLS}$ | $\theta_{Tobit}$ | $ME_{Tobit}$ | $\theta_{OLS}$ | $\theta_{Tobit}$ | $ME_{Tobit}$ |
|---|---|---|---|---|---|---|---|---|---|
| *Intercept* ($\theta_0$) | 4.133*** | 2.320** | – | 7.200*** | 6.976*** | – | 4.304** | 2.932 | – |
| *Group_member's_contribution*$_{t-1}$ ($\theta_1$) | 0.228*** | 0.344*** | 0.201*** | – | – | – | 0.291** | 0.408* | 0.260** |
| *Group_member's_received_points*$_{t-1}$ ($\theta_2$) | – | – | – | –0.080 | –0.140 | –0.089 | 0.158 | 0.193 | 0.123 |
| *N* | | | | | 288 | | | | |

***Table C3.*** Sent Disapproval Points and the Relative Size of Contribution – Extension of Table 4.

| | $\gamma_{OLS}$ | $\gamma_{Tobit}$ | $ME_{Tobit}$ |
|---|---|---|---|
| *Intercept* ($\gamma_0$) | 6.227*** | 7.831*** | – |
| *Group_member's_contribution*$_t$ ($\gamma_1$) | –0.380*** | –0.766*** | –0.433*** |
| *Sender's_positive_deviation*$_t$ ($\gamma_2$) | 0.167*** | 0.314*** | 0.182*** |
| *Sender's_absolute_negative_deviation*$_t$ ($\gamma_3$) | 0.075 | 0.128 | 0.074 |
| $1_{Rematching_t^{t+1}}$ ($\gamma_4$) | –0.536 | –1.113 | –0.644 |
| $1_{Rematching_t^{t+1}}$ × *Group_member's_contribution*$_t$ ($\gamma_5$) | 0.047 | 0.066 | 0.038 |
| $1_{Rematching_t^{t+1}}$ × *Sender's_positive_deviation*$_t$ ($\gamma_6$) | 0.010 | 0.048 | 0.028 |
| $1_{Rematching_t^{t+1}}$ × *Sender's_absolute_negative_deviation*$_t$ ($\gamma_7$) | –0.138 | –0.220 | –0.127 |
| *N* | | 576 | |

# WHEN AND WHY MATCHES ARE MORE EFFECTIVE SUBSIDIES THAN REBATES

René Bekkers

## ABSTRACT

*This paper replicates and refines the finding that subsidies for charitable contributions of a rebate type are less effective than matching subsidies. A survey based field experiment with health charities was conducted among a national sample representative of the Dutch population on key demographic characteristics. The greater effectiveness of matching subsidies found in laboratory experiments is replicated. Also some evidence is provided on why matches are more effective than rebates. Matches attract a larger pool of donors, in part because donors expect more people to make donations and "join in." Matches also increase the amount contributed among the higher educated, higher income households and larger donors. Subsidies of either type do not decrease subsequent giving in a campaign for tsunami relief. The experiment could not test whether the greater effectiveness of a matching subsidy is due to a change in the donor's attention to the benefits of a donation to the cause. This explanation should be tested in future research. The findings imply that a given budget available to subsidize charitable contributions can be*

**Replication in Experimental Economics**
**Research in Experimental Economics, Volume 18, 183–211**

**ISSN: 0193-2306/doi:10.1108/S0193-230620150000018007**

*used more effectively if the subsidy is framed in the form of a match than in the form of a rebate. Nonprofit organizations can use this insight in the design of fundraising campaigns. For governments the finding suggests that the effectiveness of current subsidies for charitable contributions can be enhanced by matching them rather than providing a deduction in the income tax, which works as a rebate.*

**Keywords:** Philanthropy; subsidies; fundraising; framing; matching; rebate

**JEL classifications:** D64; C72; C93; H41; H21; A13

# INTRODUCTION

*The Effectiveness of Matching Subsidies for Charitable Donations*

Governments throughout the world subsidize donations by offering tax deductions. Deductions for charitable donations lower the price of giving just like a rebate in the grocery store. At the 50% income tax level, a donation of 1 euro costs only 50 cents (50% rebate). Evidence from laboratory experiments with small groups of students suggests that rebates are a less effective type of subsidy for charitable donations than matching programs, which generate a mathematically equivalent price reduction (Eckel & Grossman, 2003, 2004). In a matching program, an independent third party matches donations to a charitable cause at a given rate (e.g., at 100%, for each dollar donated the third party matches a dollar). From a rational point of view, it is strange that a 100% match increases generosity more than a 50% rebate because the price of giving is the same. The practical significance of this finding is that tax deductions may be an ineffective type of subsidy.

This finding suggests that governments could more effectively subsidize donations by replacing the charitable deduction by a matching program. To inform such a policy change, however, we need answers to a series of questions. In the present paper, I will deal with four questions: (1) whether matches are really more effective than rebates; (2) why matches are more effective; (3) at which price levels and among which subgroups of the population matches are most effective; (4) whether matches undermine subsequent giving.

### *Are Matches Really More Effective than Rebates?*

There are two reasons to doubt that matches are really more effective than rebates. The first reason is that the findings of Eckel and Grossman cannot easily be generalized to everyday life situations in which people decide about giving. The experiment was conducted among a sample of university students who participated in an experiment for a show up fee, and received additional money at the beginning of the experiment that they could decide to give away. The use of "house money" in the experiment introduces a difference with situations in daily life in which people decide about giving away money that they have earned on the labor or capital market: participants may be more generous with house money (Carlsson, He, & Martinsson, 2013; Reinstein & Riener, 2012). It is unknown whether matches are also more effective than rebates among "ordinary" citizens deciding about donations outside the lab. Frey and Meier (2004) confirmed the positive effect of a matching subsidy in a field experiment with earned money among Swiss university students, but did not test for a rebate effect. An additional feature of the Eckel and Grossman studies is that the participants are selected from a pool of students. It is possible that students react differently to matching and rebate subsidies than lower educated persons. I will provide a hypothesis on this difference below.

The second reason is that the difference between matches and rebates may be an "artefact" of the procedure used by Eckel and Grossman (2003). Davis, Millner, and Reilly (2005) have argued that rebates are less effective because it was not clear enough for experimental subjects that matches and rebates are mathematically equivalent. If people are made more aware that the price of giving in a 50% rebate is the same as in a 100% match, they would give the same in both situations. I call this the *bounded rationality hypothesis*. Davis et al. (2005) obtain support for their hypothesis by confronting subjects with several exercises that illustrate how rebates and matches work out on the price of a gift before they have their subjects make the actual donation decisions. This procedure is problematic for two reasons. First, the experiment induces demand characteristics that call the validity of the results into question. The procedure makes it very clear to subjects that they are expected to make no difference between the rebate and matching conditions. The exercise sends the message "do not make mistakes!" Participants in the experiment will feel that the experimenters expect them to make rational decisions, and will act accordingly in order not to disappoint the experimenters. A second problem with the procedure used by Davis et al. (2005) is that it creates an experimental situation which is very different from everyday situations in which people decide about giving to nonprofit organizations. In daily life, donors do not complete a series of exercises before making a donation. The usual fundraising campaign confronts them with a request for a donation and only one type of subsidy.

While the procedure used by Davis et al. (2005) may not be a common situation, the bounded rationality hypothesis may still be valid. I will use a different technique to test the bounded rationality hypothesis that does not induce demand characteristics. I rely on a cognitive proficiency test to measure people's capability to understand how a subsidy affects the price of giving. Previous research in cognitive psychology has found that performance on heuristics and biases tasks is positively related to measured intelligence (Stanovich & West, 1998, 2000). Parker and Fischhoff (2005) found that resistance to framing is positively correlated with cognitive ability as measured by a vocabulary test. If the bounded rationality hypothesis is true, we should find that:

**H1.** The difference between matches and rebates is more pronounced among people with lower levels of cognitive proficiency (bounded rationality hypothesis).

Note that this hypothesis implies that matches are even more effective among persons with lower levels of education than among students, because students will on average have higher levels of cognitive proficiency.

### *Why Are Matches More Effective than Rebates?*

Eckel and Grossman (2003, 2004) suggest that matches are more effective than rebates because donors experience a match as an additional gift, while a rebate is not always viewed as such. The match buys a higher donation to the charity at no cost to the donor. They argue that offering a rebate generates a competitive orientation, while offering a matching amount generates a cooperative orientation. I call this the *framing hypothesis*. Donors experience a matching subsidy from a third party as somebody else "joining in" who might otherwise not have contributed. Matching programs raise the expectation of prospective donors that others are also donating more strongly than rebate subsidies.[1] However, the framing hypothesis has not been tested explicitly. The purpose of this paper is to conduct such a test. If the framing hypothesis is correct, we should find that:

> **H2.** The difference between matches and rebates is due to higher expectations about the donations of others under a matching subsidy (framing expectations hypothesis).

### *When Are Matches More Effective than Rebates, and for Whom?*

A third question that I address in the present study is at which price levels and among which subgroups of the population matches are most effective. Generally, the effectiveness of subsidies for giving depends on (a) the ability of the subsidy to attract new donors, who would not have donated without the subsidy; (b) the ability to increase the amount donated among donors. The *framing hypothesis* does not discriminate between these alternatives. Meier and Frey (2004) argue that the matching subsidy appeals only to those who would have donated anyway and increases their donations, but only when the proportion matched is high enough (e.g., 50%). To see this, consider the evidence from psychology that external incentives for prosocial behavior often destroy intrinsic motivation (Deci & Ryan, 1985; Frey, 1997). When donors receive a subsidy for giving, they can no longer attribute their giving to altruistic motives. In that case, donors who have altruistic motives for giving may reduce their giving because giving would endanger their altruistic self-identity. This is called the "motivational crowding-out effect." At lower subsidy levels, Meier and Frey argue, a matching program may even lower the amount donated when the motivational crowding-out effect outweighs the price effect. At sufficiently high matching rates, however, the price effect dominates the motivational crowding-out effect. This leads to the following hypotheses:

> **H3a.** Matches increase the amount donated among donors rather than the number of donors.

> **H3b.** Matches increase the amount donated among donors more strongly at higher matching rates.

The arguments put forward by Meier and Frey about motivational crowding out also point to the importance of recognizing the heterogeneity among donors. Subsidies may have different effects among donors, depending on their motivation to give. Specifically, those who have altruistic motives for giving will respond less strongly to a subsidy than those with "warm glow" motives for giving.

**H4.** Subsidies for giving increase giving less strongly among persons with more altruistic motives for philanthropy (altruistic self-identity hypothesis).

**H5.** Subsidies for giving increase giving more strongly among persons with joy of giving motives for philanthropy (joy of giving hypothesis).

The prospective design of the Giving in the Netherlands Panel Survey (GINPS) enables a detailed study of the effectiveness of subsidies for giving among subgroups of the population because data on annual donations in previous years is available, as well as a wealth of other information on donors' socio-demographic characteristics and social values. Nonprofit organizations often segment their donor populations based on prior donation levels and background characteristics. Characteristics of donors known to be related to higher giving include gender, age, level of urbanization, the level of education, income, home ownership, and religion (Bekkers & Wiepking, 2011; Wiepking & Bekkers, 2012). These characteristics are measured in the survey. I do not offer a priori hypotheses on the differential effects of matches and rebates among subgroups of the population. Donations may be subsidized even more effectively if we know whether and how the effectiveness of matches varies with the price of giving and with characteristics of donors. This knowledge can be used to target matching programs at specific groups of donors, disregarding other types of donors who would not be affected by the type of subsidy.

### *Do Subsidies Undermine Subsequent Giving?*

Finally, there is some concern about the possibility that subsidies reduce subsequent giving. If donors have received a subsidy for donations, but get no subsidy on other occasions, they may be less likely to give because the subsidy reduced their intrinsic motivation for giving. Meier (2007) found evidence for such a spoiling effect of matches on subsequent giving in the same context involving the same charity (University funds). I will test whether subsidies also reduce giving in other contexts, expecting:

**H6.** Subsidies for giving reduce giving on subsequent occasions where subsidies are not provided.

## DATA AND METHODS

### *Testing the Effectiveness of Subsidies in a Survey Field Experiment*

To study the effectiveness of matching and rebate subsidies in a natural context, a field experiment was conducted among respondents in the GINPS (Bekkers, Boonstoppel, & De Wit, 2013). The GINPS is a biennial survey on philanthropy and volunteering conducted by the Vrije Universiteit Amsterdam. The respondents receive a reward for participation in the survey. I study which proportion of these earnings the respondents give away under different types of subsidy.

*Participants*

861 respondents took part in the experiment after completing the questionnaire of the second wave of the GINPS in May 2004.[2] The GINPS uses Computer-Assisted Self Interviewing (CASI) through the World Wide Web. The first wave was completed by a random sample of 1,707 respondents from the Dutch population. The experiment was conducted after the second wave was completed. Data on personal characteristics of the participants collected in both waves 1 and 2 (2002 and 2004) are used in the analyses. The respondents are a representative sample of the Dutch population with regard to age, gender, level of education and region of residence.[3]

*Procedure*

By participating in the GINPS, respondents earned a reward, proportional to the amount of time it took them to complete the survey. The reward was given in the form of "points," which corresponded to an amount in Euros. The monetary value of the "points" earned was displayed after the respondents completed the survey. Then, the respondents were given the following choice (henceforth called the "donation decision") between two alternative options. The first option consisted of an exchange of the points into Air Miles or vouchers for a well-known chain of shops for domestic articles, music and video products. The second option consisted of an exchange of earnings into a donation to one of three charitable causes: "Médecins Sans Frontières" (MSF; in the US known as Doctors Without Borders), providing emergency health care in developing countries, the Aids Fund (fighting HIV/Aids, mainly through research), or the Queen Wilhelmina Fund (fighting cancer, mainly through medical research).[4]

Completing the entire GINPS-questionnaire took the respondents about 45 minutes. The mean amount earned was €11.08 ($13.35, exchange rate $1 = €0.8369, May 16, 2004). Because some respondents took more time to complete the survey than others, the amount of money involved in the donation decision differs between participants. The lowest earnings were 6.30 and the highest 15.75. Earnings were included as a covariate in all regressions because they are a measure of the strength of temptation to keep the money.[5] The pass rate is the amount donated by earnings.

As in the "dictator game" used by Eckel and Grossman (2003) the respondents could decide to give away any desired proportion of the money. Unlike Eckel and Grossman's study, the participants in the present study did not decide about a windfall endowment they received at the beginning of the experiment, but decided about money they earned by participating in the survey. Cherry, Frykblom, and Shogren (2002) show that subjects playing a dictator game with earned income are much less generous than subjects playing with a windfall endowment received from experimenters. In a prior experiment I replicated this finding in a national sample of the Dutch population among the respondents of the first wave of the GINPS (Bekkers, 2007). Having respondents decide about donations with earnings is a more valid procedure than having them decide after providing a windfall endowment because actual donations are also made with earned rather than endowed wealth.

After the actual donation decision, the respondents are asked how the average Dutch citizen would respond in the same condition. The respondents reported whether they thought the average person would have donated, and if so, how much. The expectation that others would donate was strongly related to actual donation decisions ($r$ = .465, $p$ < .000; Chi Square 185.89, df = 1, $p$ < 000). The expected pass rate was also strongly related to the actual pass rate ($r$ = .513, $p$ < .000).[6]

Nine months after the experiment, 704 of the 861 participants in the experiment (81.8%) responded to a survey on tsunami relief giving in the aftermath of the disaster in South East Asia on December 26, 2004. The total amount donated to tsunami relief efforts served as a criterion variable for testing the hypothesis that receiving a subsidy would decrease subsequent giving. Attrition was randomly distributed over experimental conditions and was not related to earnings or donations in the experiment.

### *Experimental Design*

Respondents are placed randomly in one of the five conditions in the experiment (see Table 1). In the base line condition, participants did not

***Table 1.*** Experimental Design.

| Condition | 1. Base Line | 2. 100% Match | 3. 50% Match | 4. 50% Rebate | 5. 33% Rebate |
|---|---|---|---|---|---|
| Match | No | 100% | 50% | No | No |
| Rebate | No | No | No | 50% | 33% |
| Price of giving €1 | €1.00 | €0.50 | €0.67 | €0.50 | €0.67 |
| *N* | 165 | 173 | 184 | 180 | 159 |

get a subsidy for their gifts. Conditions 2–5 vary the price of giving as in previous research (Davis et al., 2005). In condition 2, donations by the respondents are matched one to one by a third party (TNS/NIPO, the opinion poll research institute conducting the survey). In this condition, a donation of €4 effectively turns into a donation of €8, lowering the price of giving with 50%. Mathematically, the price of giving in condition 2 is the same as in condition 4, where respondents get a 50% rebate. The matching condition parallels the "Buy one, get one free" strategy that is used to increase sales in supermarkets. The rebate condition parallels the "Now 50% off!" strategy that is also often used to reach the same goal. Conditions 3 and 5 are matching and rebate conditions at the .67 price level.

### *Measures*

A measure of *cognitive proficiency* was included to test the bounded rationality hypothesis. Cognitive proficiency was measured with a vocabulary test (Gesthuizen & Kraaykamp, 2002). Respondents read 13 words in Dutch. For each word, respondents were asked to pick a synonym from a set of five other words. Strictly speaking, this measure taps verbal ability, but prior research indicates that measures of verbal ability are highly correlated with other measures of learned or "crystallized" intelligence (Alwin, 1991).

Two measures of motives for philanthropy were included to test the motivational crowd out hypothesis. All items were Likert-scales consisting of items on a five-point scale (disagree completely – agree completely). *Altruistic values* were measured with a scale consisting of 8 items (alpha = .81) adapted from Lindeman (1995; sample item: "I don't like to spend time doing things for others").[7] *Joy of giving* was measured with three items (alpha = .78): "Giving to charity gives me a positive feeling," "Giving to charity makes me happy," and "Giving to charity makes me more happy than buying something for myself".

Included in the present study are measures for *gender* (female = 1), *age* (in years), the *level of education* (7 categories, ranging from primary school only to university degree), *gross yearly household income* (midpoint values were used of 24 categories, ranging from €2,000 to €300,000; higher incomes were truncated), *home ownership* (yes = 1), *religious affiliation* (dummy variables for Catholic, Reformed Protestant, Rereformed Protestant, or other), *frequency of church attendance* (the number of church visits per year), and the *amount donated to nonprofit organizations in 2001* (total amount in euros).

Because the charities the participants could donate to are all in the health sector, I included measures of *subjective health* (ranging from 1 – poor – to 5 – very well) and *health problems in one's social network* (the number of chronic diseases among friends and family in the past two years).

Five measures for the relationship with charities were included as well: *charitable confidence* ("How much confidence do you have in charities?," ranging from 1 – none at all – to 5 – very much); the number of *personal* and *impersonal solicitations* for charitable contributions in the two weeks preceding the survey; whether the respondents *know the "seal of approval"* for fundraising nonprofit organizations ("CBF-keur," Bekkers, 2003); and *social pressure* to volunteer for nonprofit organizations, indicated by agreement on a 1 (disagree completely) to 5 (agree completely) scale with the statement "In my social environment, it is obvious to volunteer" (Bekkers & Schuyt, 2008).

# RESULTS

## *Basic Findings: Replicating the Matching Effect*

Table 2 shows the distribution of donation decisions in the experiment. Compared to the rebate, the proportion of respondents keeping the reward

***Table 2.*** Donations in Experimental Conditions ($n$ = 861).

| Price | Frame ($n$) | Keep All | Donate Something | Donate 1–99% | Donate 100% |
|---|---|---|---|---|---|
| 1.00 | No framing (165) | 64.8 | 35.2 | 32.1 | 3.0 |
| 0.67 | Rebate (159) | 62.9 | 37.1 | 35.8 | 1.3 |
| 0.67 | Match (184) | 48.9 | 51.1 | 46.2 | 4.9 |
| 0.50 | Rebate (180) | 52.2 | 47.8 | 44.4 | 3.3 |
| 0.50 | Match (173) | 42.8 | 57.2 | 52.0 | 5.2 |

completely (vs. giving away at least something) is significantly higher in the match and rebate condition at the ⅔ price level (Chi Square = 6.747 (df = 1), $p < 009$). The difference at the ½ price level is marginally significant (Chi Square = 3.157 (df = 1), $p < 076$). This result is not consistent with the hypothesis of Frey and Meier (2004) that matches would be more effective at a lower price of giving.

Table 3 shows the amounts donated and pass rates (the proportion of the earnings donated before subsidy) in the experiment. Both at the .67 and the .50 price level, matches generate higher amounts contributed than rebates (Anova $F$-statistic = 5.600 (df = 1), $p < 019$; and $F = 4.556$ (df = 1), $p < 034$, respectively). Among donors, however, there are no significant differences in the amount donated (for price = 0.67: $F = 0.083$ (df = 1), $p < 773$; for price = 0.50: $F = 1.367$ (df = 1), $p < 244$). These results differ markedly from those obtained by Eckel and Grossman (2003) and Davis et al. (2005), who found that participants tended to contribute the same proportion of their endowment to the charity (about 50% and 40%, respectively), regardless of the type of subsidy. In the present experiment, matches did increase the pass rate.

Among donors, however, pass rates are between 40% and 50% in all conditions. Inspection of the distribution of pass rates among those who donated (see Fig. 1) reveal that there is a strong preference for giving exactly 50% of the earnings to charity. This result holds in all conditions (see Appendix for separate figures for each condition).

So including the zeros, the difference between matches and rebates is significant, but excluding the zeros it is not. This suggests that the match is mainly effective because it attracts more donors, but not larger donations. I will test this below using regression models.

***Table 3.*** Amounts Donated and Pass Rates in Experimental Conditions ($n = 861$).

| Price | Frame ($n$) | Among All Respondents | | Among Donors Only | |
|---|---|---|---|---|---|
| | | Amount donated | Pass rate | Amount donated | Pass rate |
| 1.00 | No framing (165) | 1.71 | 15.36 | 4.86 | 43.70 |
| 0.67 | Rebate (159) | 1.88 | 17.17 | 5.06 | 46.26 |
| 0.67 | Match (184) | 2.64 | 24.19 | 5.18 | 47.36 |
| 0.50 | Rebate (180) | 2.46 | 22.47 | 5.15 | 47.03 |
| 0.50 | Match (173) | 3.19 | 29.14 | 5.58 | 50.92 |

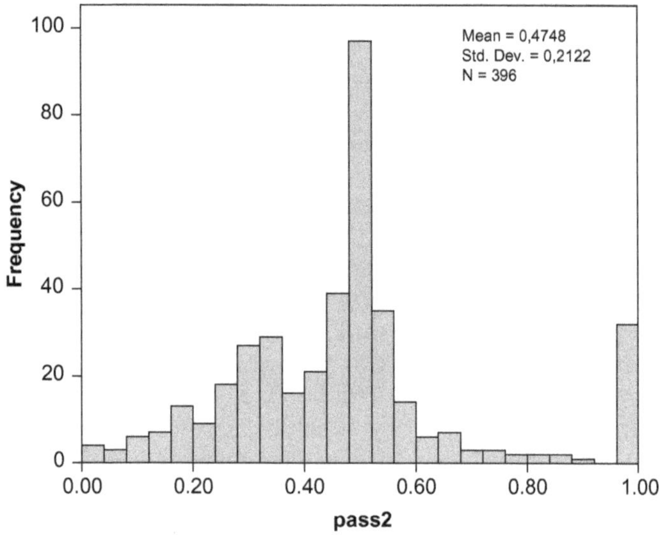


*Fig. 1.* Distribution of Pass Rates (All Conditions Pooled).

***Table 4.*** Donations Expected from Others in Experimental Conditions ($n = 861$).

| Price | Frame | Keep All | Donate 1–99% | Donate 100% | Amount Donated | € Among Donors | Pass Rate | Pass Rate (Donors) |
|---|---|---|---|---|---|---|---|---|
| 1.00 | No frame | 28.5 | 70.3 | 1.2 | 3.27 | 4.58 | 29.16 | 40.78 |
| 0.67 | Rebate | 31.4 | 65.4 | 3.1 | 3.51 | 5.12 | 31.61 | 46.10 |
| 0.67 | Match | 19.0 | 78.8 | 2.2 | 3.59 | 4.44 | 33.28 | 41.10 |
| 0.50 | Rebate | 23.9 | 71.7 | 4.4 | 3.81 | 5.01 | 35.21 | 46.26 |
| 0.50 | Match | 18.5 | 79.2 | 2.3 | 3.84 | 4.71 | 34.94 | 42.87 |

Recall that the framing hypothesis predicts that expectations about the contributions of others mediate the influence of the subsidy scheme. If that hypothesis is correct, we should find that the participants expect higher contributions from other participants in matching conditions than in rebate conditions. Table 4 gives little support for this prediction.

While at the .67 price level participants indeed expected more others to donate at least some of their earnings (Chi Square = 7.064, df = 1, $p < .008$), at the .5 price level they did not (Chi Square = 1.533, df = 1, $p < .216$). In addition, they did not expect higher over all contributions in matching conditions ($F$-values in Anovas at price = .5, $F$ = .080, df = 1, $p$ < .778; at price .67, $F$ = .011, df = 1, $p$ < .918). This is due to the fact that those who expected that others would give away at least some of their earnings, expected that a rebate would increase the amount donated more strongly than a rebate (at least at price .67, $F$ = 6.017, df = 1, $p$ < .015; at price = .5, $F$ = 1.272, df = 1, $p$ < .280). In other words, the participants expected the match to be more effective in getting people to give than a rebate, but only at a lower subsidy, while at the same time they expected the contributions of others to be lower. Thus, the respondents thought that other people would be rather insensitive to price changes as well as framing effects. This makes it unlikely that expectations about the behavior of others mediate the influence of matching subsidies.

### *Multivariate Regression Models: Testing Hypotheses*

Table 5 reports further evidence on the effects of matches and rebates on giving using a Heckman two-stage regression model (Heckman, 1979; for an application to charitable giving, see Smith, Kehoe, & Cremer, 1995). Stage 1 of this model relates the type of subsidy to the likelihood that people gave anything. Stage 2 of the model relates the type of subsidy to the proportion of the earnings donated, conditioning on the effect of matches on whether the respondents give at all. The results in model 1 reveal that matches increase the likelihood of a donation more strongly than rebates as well as the pass rate. Of course, a lower price also increases giving. Earnings do not affect giving in either stage.

**H1.** No support for bounded rationality hypothesis

Model 2 reveals that matches are equally effective in promoting the likelihood of giving as well as the amount donated among people with varying levels of cognitive proficiency. This finding contradicts the bounded rationality hypothesis. The main effect of cognitive proficiency is positive in both stages: those with higher levels of cognitive proficiency are somewhat more likely to give, but most importantly, if they give, they give a higher proportion of their earnings.[8]

**H2.** Limited support for the framing expectations hypothesis

Model 3 reveals that expectations mediate only a small part of the influence of a matching subsidy, as suggested by Table 4. The expectation that

***Table 5.*** Heckman Two Stage Regression of Donations Testing the Bounded Rationality Hypothesis and the Framing Hypothesis ($n = 861$).

| | Model 1 | | Model 2 | | Model 3 | |
|---|---|---|---|---|---|---|
| | Whether | Rate | Whether | Rate | Whether | Rate |
| Match | .275** | .080** | .359* | .105* | .257* | .093*** |
| | (.094) | (.028) | (.155) | (.048) | (.103) | (.023) |
| Price | −1.162* | −.370* | −1.103* | −.339* | −1.174(*) | −.293* |
| | (.561) | (.168) | (.564) | (.167) | (.612) | (.137) |
| Earnings | −.039 | −.014 | −.032 | −.016 | .032 | .001 |
| | (.036) | (.010) | (.036) | (.010) | (.038) | (.008) |
| Cognitive proficiency | | | .240(*) | .118*** | | |
| | | | .134 | (.045) | | |
| Match × cognitive proficiency | | | −.129 | −.040 | | |
| | | | (.178) | (.056) | | |
| Expectation that others donate | | | | | 1.291*** | −.006 |
| | | | | | (.206) | (.060) |
| Expected amount donated | | | | | 1.225*** | .752*** |
| | | | | | (.274) | (.063) |
| *N* (censored, uncensored) | 358 | 338 | 358 | 338 | 358 | 338 |
| Chi Square (df) | 14.01*** (3) | | 26.03*** (5) | | 177.78*** (5) | |
| Independence LR test | 29.71*** | | 30.55*** | | 38.00*** | |

***$p < .001$; **$p < .01$; *$p < .05$; (*)$p < .10$.

others will donate strongly increases the likelihood that people will give, and also increases the pass rate. However, controlling for expectations about the donations of others does little to reduce the difference between matching and rebate subsidies. The effect of a matching subsidy on the likelihood of donating weakens a little (from .275, $p < .003$ to .257, $p < .013$), but remains significant. A more positive expectation about the amount that others will contribute strongly increases the amount contributed, but not the likelihood of making a contribution at all.[9]

**H3a.** Matches increase the number of donors rather than the amount donated

The findings in Table 2 suggested that matches primarily increase donations because they attract more donors, but not larger donations among donors. However, the findings in Table 5 showed that matches increased both the likelihood of donating and the amount donated among donors.

**H3b.** Increasing marginal effects of price reductions only for rebates

The findings in Table 2 also suggested that 100% matches were not more effective than 50% matches, and Table 6 shows this is indeed the case in a formal test. For rebates, however, the expected pattern is found: 50% rebates were more effective than 33% rebates. The result for matches is in contrast with the prediction from motivation theory in hypothesis 3b. The result also stands in contrast with findings reported by Meier and Frey (2004), who found that matches became more effective at higher levels because the amount donated among donors increased.

**H4 and H5.** Motivational crowding in effects rather than crowding out effects

Table 7 shows effects of joy of giving and altruistic values in matching conditions in order to test the hypotheses on crowing out of intrinsic motivation. The results reject the hypotheses based on the theory of motivational crowd out. In contrast to the joy of giving-hypothesis, the effect of joy of giving was not stronger in the two subsidy conditions than in the no-subsidy condition. Moreover, joy of giving did not promote giving in the rebate conditions, while it did promote giving in the base line condition and the matching condition. The effect of altruistic values also did not vary with subsidies as predicted by the altruistic self-identity hypothesis. The effect of altruistic values was even stronger in the rebate condition than in the no-subsidy condition. When offered a rebate, respondents with stronger

***Table 6.*** Tobit Regression of Donations Testing Effects of Matches at Varying Price Levels (n = 696).

| | Model 1 | | Model 2 | |
|---|---|---|---|---|
| | Coeff. | S.E. | Coeff. | S.E. |
| Match | .138 | .042*** | | |
| Price | −.604 | .255* | | |
| 50% match | | | −.085 | .058 |
| 50% rebate | | | −.122 | .058* |
| 33% rebate | | | −.241 | .062*** |
| Earnings | −.020 | .016 | −.020 | .016 |
| Constant | .530 | .223* | .354 | .183(*) |
| *N* (censored, uncensored) | 358 | 338 | 358 | 338 |
| Chi Square (df) | 17.54*** (3) | | 17.69*** (4) | |
| Pseudo *R* Square | .0172*** | | .0174*** | |

Participants in base line condition excluded. Reference category: 100% match condition.
A Tobit specification is used instead of Heckman model due to convergence problems.
***$p < .001$; **$p < .01$; *$p < .05$; (*)$p < .10$.

***Table 7.*** Effects of Joy of Giving and Other Regarding Preferences ($n = 861$).

| | No Subsidy | | Match | | Rebate | |
|---|---|---|---|---|---|---|
| | Whether | Pass rate | Whether | Pass rate | Whether | Pass rate |
| Joy of giving | .197(*) | −.047 | .200** | .022 | .047 | −.017 |
| | (.117) | (.041) | (.067) | (.018) | (.076) | (.022) |
| Altruistic values | .068 | .018 | .082 | .014 | .152* | .018 |
| | (.111) | (.033) | (.072) | (.020) | (.076) | (.031) |
| *n* (censored, uncens) | 107 | 58 | 164 | 193 | 194 | 145 |
| Chi Square (df) | 2.12 (3) | | 6.07 (3) | | 1.68 (3) | |
| Independence test | 0.00 | | 15.34*** | | 0.00 | |

Price and earnings also included; effects not shown.
***$p < .001$; **$p < .01$; *$p < .05$; (*)$p < .10$.

altruistic values were more likely to give away at least something, while altruistic values did not promote giving when offered a matching subsidy or no subsidy at all. This pattern clearly runs counter to the motivational crowding out hypothesis.

### *Differential Effects of Matches among Specific Groups*

Table 8 reports a multinomial logistic regression analysis of donations on socio-demographic characteristics of the respondents in the experiment. The model allows for a test of the influence of matches among specific groups on making donations of small parts of the earnings, about half of the earnings, or more. A multinomial specification allows for a more detailed analysis of the effect of matches on different levels of generosity.

Matches are equally effective among different age groups, among persons with different religious affiliations,[10] among home owners and those renting a home, and among residents of metropolitan areas and towns with different sizes. However, the effect of matches did vary with gender, level of education, income, and prior donations.

While females are less likely to donate a small amount than males, matches reverse the pattern: females are more responsive to matches than males.[11] Higher educated persons are more responsive to matches than lower educated persons, but this trend reverses at the highest level of education. Those who hold a bachelor degree actually *reduce* their giving when they receive a matching subsidy. Large donors in 2001 were more likely to donate half of their earnings in the experiment than small donors, regardless of the type of

***Table 8.*** Multinomial Logit of Donations on Personal Characteristics.

| | 1–45% | | 45–55% | | 55–100% | |
|---|---|---|---|---|---|---|
| | Coeff. | S.E. | Coeff. | S.E. | Coeff. | S.E. |
| Female | −.636 | .302* | −.070 | .330 | .166 | .403 |
| Female × match | 1.101 | .433* | .189 | .435 | −.404 | .539 |
| Age | −.001 | .011 | .007 | .012 | .017 | .013 |
| Age × match | −.011 | .015 | −.001 | .015 | −.003 | .017 |
| Town size | −.000 | .001 | −.000 | .001 | −.000 | .001 |
| Town size × match | −.000 | .002 | −.000 | .001 | .001 | .002 |
| Level of education | −.042 | .149 | .039 | .162 | .020 | .212 |
| Education × match | .472 | .218* | .036 | .222 | .386 | .283 |
| BA degree or higher | −.365 | .667 | −.474 | .689 | .843 | .860 |
| Degree × match | −1.793 | .959(*) | −.042 | .955 | −2.120 | 1.161(*) |
| Donations in 2001 | −.009 | .049 | .142 | .066* | .033 | .069 |
| Donations × match | .137 | .082(*) | .002 | .087 | −.010 | .091 |
| Income (ln) | −.589 | .300* | −.106 | .350 | −.674 | .390(*) |
| Income × match | .071 | .428 | .032 | .444 | 1.113 | .522* |
| Own home | .052 | .340 | −.237 | .369 | .772 | .496 |
| Own home × match | .114 | .488 | .022 | .480 | −.469 | .628 |
| Catholic | .447 | .397 | −.053 | .472 | −.228 | .605 |
| Catholic × match | .604 | .567 | .116 | .635 | .522 | .782 |
| Reformed | .451 | .471 | .318 | .527 | .542 | .581 |
| Reformed × match | −.325 | .659 | −.274 | .681 | −.819 | .857 |
| Rereformed | .349 | .599 | .198 | .583 | −.523 | .840 |
| Rereformed × match | −1.207 | .869 | −.421 | .758 | −.322 | 1.042 |
| Other religion | −.079 | .599 | −.701 | .658 | .376 | .623 |
| Price | −1.042 | 1.295 | −4.067 | 1.309** | −1.501 | 1.590 |
| Match | −1.152 | 1.878 | .480 | 1.929 | −3.680 | 2.309 |
| Earnings | .049 | .092 | −.100 | .090 | −.220 | .106* |
| Constant | 1.434 | 1.711 | 2.811 | 1.871 | 1.477 | 2.098 |

Pseudo $R$ Square .0718; LR Chi Square 121.01 (df = 78), $p < .0013$.
$^{***}p < .001$; $^{**}p < .01$; $^{*}p < .05$; $^{(*)}p < .10$.

subsidy. Matches were more effectively promoting small donations among large donors. While persons in higher income households were less likely to donate both a small and a major part of their earnings, matches increased the likelihood of donating more than half of the earnings at higher income levels.

Because the three charities that benefited from donations in the experiment were health charities, experiences with health problems in one's own life or in one's close social circle are relevant. Those who know people with health problems may be more likely to give in the experiment because they identify more easily with the (ultimate) recipients of their gifts. Previous research shows that giving increases with one's own health (Bekkers, 2008), and with knowing people with health problems (Bekkers, 2008; Meijer, Bekkers, & Schuyt, 2005). The results in Table 9 reveal that matches were less effective than rebates in generating small donations among those in better health, and less effective in generating large donations among those with more health problems in their social network. Matches were also less effective than rebates in generating small donations among those who discuss giving to charities more often. It seems that matches reduce giving among those who are more able to identify with recipients of their gifts. Table 9 also shows that the effect of matches is no different among those with different levels of confidence in charities, exposure to fundraising attempts, and awareness of the monitoring system for fundraising organizations.

### *Matches Reduce Giving from Identification Motives*

Table 9 reports a multinomial logistic regression analysis of donations on attitudes towards fundraising nonprofit organizations and experiences with health problems.

***Table 9.*** Multinomial Logit of Donations on Attitudes to Giving.

| | 1–45% | | 45–55% | | 55–100% | |
|---|---|---|---|---|---|---|
| | Coeff. | S.E. | Coeff. | S.E. | Coeff. | S.E. |
| Confidence | .270 | .280 | .802 | .246*** | .513 | .278(*) |
| Confidence × match | .183 | .290 | −.412 | .312 | .239 | .380 |
| Solicitations | −.095 | .110 | −.101 | .122 | −.092 | .141 |
| Solicitations × match | .058 | .157 | −.118 | .169 | −.013 | .200 |
| Social pressure | .071 | .161 | .516 | .184** | .443 | .215* |
| Pressure × match | .209 | .228 | −.335 | .237 | −.311 | .286 |
| Know seal | .152 | .438 | −.401 | .351 | −.022 | .402 |
| Know seal × match | .152 | .438 | .623 | .456 | .507 | .545 |
| Talk about giving | .466 | .288 | −.169 | .319 | .142 | .378 |
| Talk × match | −.827 | .384* | .007 | .397 | −.437 | .489 |
| Subjective health | .414 | .169* | .086 | .199 | .085 | .229 |
| Health × match | −.416 | .235(*) | −.189 | .253 | .180 | .301 |
| Health problems | .008 | .100 | .068 | .115 | .161 | .126 |
| Problems × match | .132 | .138 | −.022 | .149 | −.318 | .190(*) |
| Price | −.198 | 1.270 | −3.679 | 1.310** | −2.006 | 1.588 |
| Match | 1.420 | 1.430 | 3.567 | 1.541* | 1.047 | 1.864 |
| Earnings | −.012 | .084 | −.148 | .085(*) | −.235 | .100* |
| Constant | −3.798 | 1.460* | −1.490 | 1.614 | −1.610 | 1.870 |

Pseudo $R$ Square .0587; LR Chi Square 99.04 (df = 51), $p < .0001$.
***$p < .001$; **$p < .01$; *$p < .05$; (*)$p < .10$.

**H6:** Subsidies promote Tsunami Relief giving

Table 10 shows how the offer of a subsidy for donations in the experiment is related to subsequent giving, that is, giving in response to the tsunami disaster. In contrast to the result reported by Meier (2007), receiving a subsidy in the experiment did not decrease subsequent giving to tsunami relief efforts. Moreover, respondents who were placed in the 33% rebate condition in the experiment even gave more than respondents in the base line condition, who had received no subsidy.

Model 2 adds interactions between experimental conditions and behavior, because Meier (2007) found that the detrimental effect of matches was stronger in a low subsidy condition (i.e., 50%) than in a high subsidy condition (i.e., 25%). The present analysis shows very different results: rebates and matches did not decrease giving more strongly among those who had made donations in the experiment. If anything, those who had donated in

***Table 10.*** Tobit Regression of Amount Donated in Tsunami Relief Campaign.

| | Model 1 | | Model 2 | | Model 3 | |
|---|---|---|---|---|---|---|
| | Coeff. | S.E. | Coeff. | S.E. | Coeff. | S.E. |
| Match 100% | .021 | .222 | .059 | .300 | −.460 | .344 |
| Match 50% | −.042 | .218 | .019 | .277 | −.452 | .315 |
| Rebate 50% | .101 | .213 | −.119 | .258 | −.366 | .296 |
| Rebate 33% | .512 | .223* | .604 | .261* | .359 | .292 |
| Earnings | .382 | .057*** | .331 | .057*** | .332 | .057*** |
| Match 100% × donated | | | −.044 | .362 | −.241 | .372 |
| Match 50% × donated | | | .027 | .345 | −.220 | .363 |
| Rebate 50% × donated | | | .619 | .338(*) | .577 | .357(*) |
| Rebate 33% × donated | | | −.264 | .367 | −.312 | .392 |
| Joy of giving | | | .015 | .133 | .013 | .132 |
| Joy of giving × match | | | .475 | .186* | .515 | .185** |
| Joy of giving × rebate | | | −.006 | .186 | −.005 | .184 |
| Altruistic values | | | .423 | .136** | .426 | .134** |
| Altruistic values × match | | | −.320 | .192(*) | −.346 | .191(*) |
| Altruistic values × rebate | | | −.065 | .190 | −.064 | .189 |
| Expected donations | | | | | −.111 | .045* |
| Expected × match | | | | | .220 | .069*** |
| Expected × rebate | | | | | .127 | .067(*) |
| Chi Square (df) | 53.10*** (5) | | 94.70*** (15) | | 105.30*** (18) | |
| Pseudo $R$ Square | .0164 | | .0292 | | .0324 | |

***$p<.001$; **$p<.01$; *$p<.05$; (*)$p<.10$.

response to a 50% rebate gave more in response to the tsunami relief campaign.

Model 2 also includes interactions between the type of subsidy and joy of giving and altruistic values. If motivational crowd out took place after the experiment, it should reduce giving to the tsunami campaign among those who give for altruistic reasons, while it should increase giving among those who give for the "warm glow." The analysis provides some support for motivational crowd out among those who were offered a match, but not among those who were offered a rebate. Matches increase giving among those who report a stronger joy of giving. This is important because joy of giving in itself does not promote tsunami relief giving. For altruistic values, the pattern is the reverse: persons with stronger altruistic values give more to tsunami relief, but not when they had been offered a matching subsidy in the experiment.

Finally, model 3 adds interactions between the type of subsidy and the generosity expected of others in the experiment. The analysis reveals that subsidies – and matches more so than rebates – promoted giving more strongly among those who held more positive expectations about the generosity of others. Including these interactions, the offer of a 33% rebate in the experiment no longer had a significant effect on tsunami relief giving.[12] These results indicate that subsidies promote subsequent giving only among those who held more positive expectations about other people's generosity. These results square nicely with the finding of Shang and Croson (2009) that more positive expectations about other's generosity not only increase current giving, but also giving on subsequent occasions when subsidies are no longer provided. Shang and Croson experimentally induced expectations, providing potential donors with information about the behavior of others.

## CONCLUSION AND DISCUSSION

This paper investigated four questions: (1) whether the same pattern of results obtained in previous research with lab experiments among small samples of university students also holds in a natural context; (2) whether the greater effectiveness of matches holds for all types of donors; (3) when and why matches are more effective than rebates; (4) whether subsidies undermine subsequent giving.

The answer to the first question is clearly "yes." The stronger effectiveness of matching subsidies also occurs in a large sample of the Dutch population ($n = 1{,}030$) in a natural context. The participants in the present study decided about the donation of a reward earned previously by filling out the questionnaire of the GINPS. By having the participants decide about earned income, I improve the ecological validity of the "dictator game" used in experimental economics (Cherry et al., 2002). The participants could donate their earnings in any desired proportion to one of three popular health charities. In a baseline condition, donations were not subsidized. In the rebate condition, an independent third party (the polling institute) offered the participants a rebate for donations. In the matching condition, the third party offered the participants a matching donation to the charity of their choice. I tested rebates against matches at two price levels, generating a 50% and 33% price reduction for donations. At both price levels, I find that matches increase donations more than rebates, but the effect is more pronounced at a price reduction level of 50%.

The results of the experiment are inconsistent with the *bounded rationality hypothesis* offered by Davis et al. (2005), who argue that differential framing is not the reason for the observed difference but a failure to see the equivalence of the two conditions. I find no evidence for *Hypothesis 1: the difference between matches and rebates is more pronounced among people with lower levels of cognitive proficiency*. More intelligent participants gave more in matching as well as rebate conditions. While cognitive ability is positively related to performance in heuristics and bias tasks, this does not hold for each and every task. Apparently, the framing effect of matches on charitable donations is largely independent of cognitive proficiency.

The answer to the second question is less clear. I found that matches are more effective than rebates partly because donors expect that matches will attract a larger number of donations by others in the matching conditions. The results give partial support for *Hypothesis 2: the difference between matches and rebates is due to higher expectations about the donations of others under a matching subsidy*. This finding is consistent with the *framing hypothesis* that Eckel and Grossman offered as an explanation for their findings. They argue that offering a rebate generates a competitive orientation, while offering a matching amount generates a cooperative orientation. Donors experience a matching subsidy from a third party as somebody else "joining in" who might otherwise not have contributed. Matching programs raise the expectation of prospective donors that others are also donating more strongly than rebate subsidies. In addition, donors may perceive the donations of others as a signal that giving is the appropriate course of action. On the other hand, donors also expect smaller sized donations from others. That is why expectations do not fully explain the superiority of matches over rebates. Clearly, further research is needed to understand why matches generate higher donations than rebates.

In the current experiment, matches are more effective than rebates mainly because they attract a larger pool of donors, not because they generate higher amounts donated. This result is in contrast with *Hypothesis 3: the effect of matches increases with the level of subsidy because matches increase the amount donated among donors.*

When donations are matched by a third party, a group of donors makes donations who would have elected not to donate when donations are subsidized with a (mathematically equivalent) rebate. However, among specific groups of the Dutch population, matches also increase the amount contributed more strongly than rebates. I find that matches are particularly effective in increasing the amount donated among the higher educated, among higher income households, and among households making larger donations to charities. These findings support the argument that the use of matches as an alternative means for subsidizing charitable donations will generate higher contributions. I also find evidence that a rebate is less effective when the price reduction is low (33% instead of 50%). In contrast to the results obtained by Meier and Frey (2004), the relative effectiveness of a matching program does not vary with the size of the price reduction. In contrast with the *altruistic self-identity hypothesis (hypothesis 4)* and *the joy of giving hypothesis (hypothesis 5)*, subsidies did not undermine intrinsic motivation for giving.

The answer to the fourth question is clearly "no." The results clearly rejected *Hypothesis 6: subsidies for giving reduce giving on subsequent occasions where subsidies are not provided.* Matching subsidies offered in the experiment did not decrease giving nine months later in response to the tsunami disaster. A small rebate (33%) even increased tsunami relief giving nine months after the experiment.

In sum, matches are more effective than rebates among three groups in the population that are very important for philanthropy: large donors, the higher educated, and higher incomes. Households with higher levels of education and household income give more than household with lower levels of education and income. Controlling for the amount donated in 2001, matches are still more effective than rebates among these groups. This is an important result: it suggests that matches can be used to increase giving among these groups even further.

Finally, there could be alternative reasons why matches are more effective than rebates that have little to do with the expectations about the behavior of others. One reason could be that matches benefit a nonprofit organization directly, while rebates do so indirectly by transferring money to the donor making a contribution. Thus, the match makes it more evident that the third party is also sponsoring the cause. This may have the psychological effect of focusing the donor's attention on the benefits of a donation to the cause. A rebate, in contrast, focuses the donor's attention on the costs of the gift to oneself. As the present experiment did not allow for a test of this hypothesis, it should be tested in future experiments.[13]

# NOTES

1. In addition, donors may perceive the donations of others as a signal that giving is the appropriate course of action. Potters, Sefton, and Vesterlund (2004) show that donations by others in a sequential setting may "set the example." Because the subjects in the current experiment decided independently and participants were not given any information about the behavior of others in the experiment, the leadership phenomenon cannot account for the effect of the matching subsidy.

2. For reasons unknown, 216 respondents were unable to participate in the experiment. This was a random group. None of the predictors of giving in the analyses was related to participating in the experiment. An additional random group of 169 respondents were excluded from the analyses below because they made a different type of decision that is difficult to compare with previous studies (an "all-or-nothing" decision, either keeping the earnings or giving them away completely; see Bekkers, 2007).

3. Respondents were drawn randomly from a pool of 72,000 respondents who regularly participate in surveys through the Internet. Persons in this pool were originally drawn randomly from the population register, and were invited in a letter to participate in surveys. Prospective participants who did not possess a suitable personal computer with Internet connection were provided with one. Data for the present study were collected through the Internet. Because there was no interviewer present when the respondents completed the GINPS-questionnaire and decided about donating their earnings, experimenter demands are unlikely to play a role in donation decisions. In drawing the sample, special care was taken to avoid sample bias with regard to Internet use by stratification with regard to age, gender, and geographic region. More details on design, sampling, and questionnaires are given in Bekkers et al. (2013).

4. Donations may have been constrained by the fact that only one of three charities could be chosen. However, the three charities are well known in the Netherlands. In terms of fundraising income, the Queen Wilhelmina Fund is one of the largest charities in the country (Central Bureau of Fundraising, 2003). In addition, two of the three charities are active in the medical research sector, which is a very common cause for donations in the Netherlands because many health charities organize door-to-door collections with high response rates (Schuyt, 2003). In the first wave of the GINPS, almost 7 out of 10 (69.4%) of the respondents reported donations by their household to health causes in the past year.

5. Donation decisions were automatically stored in a secure electronic database on the Internet, as were the responses in the GINPS-questionnaire. The complete survey session was held without interference by a human interviewer or experimenter. After the donation decision was made, an automatic e-mail message was sent to the respondents confirming their decision (e.g., "We have transferred an amount of € xxx to your bank account number xxx"). This procedure bears similarity to a single blind-procedure in ordinary dictator games. It is not double blind because respondents may believe that their responses and donation decisions can be traced by data analysts by matching the data obtained in the survey session with the e-mail address used for soliciting participation.

6. Interestingly, the respondents expected the average citizen to be more generous than they were themselves. Among those who kept their earnings, 58% expected the average citizen to give away at least something.

7. Full details of all items used in the survey are provided in the GINPS user manual (Bekkers et al., 2013).

8. A multinomial logit analysis distinguishing three categories of pass rates (1–45%, 45–55%, and 55–100%) revealed that the interactive effect of cognitive proficiency and subsidies was most pronounced on giving 55% or more.

9. The effect of earnings changes signs in model 3 because expectations interact with earnings. Additional analyses (available from the author) reveal that a more generosity expected from others affected own generosity more strongly when the price of giving was lower. The effects of expectations did not interact with receiving a subsidy or the type of subsidy. A multinomial logit analysis revealed that the effects of expectations were strongest on making a large donation (pass rate > 55%). It should be noted that expected contributions were not manipulated in the present study. They may have been shaped by a "justification effect": participants may expect others to do as they do in order not to make their own behavior appear as "deviant" (Meier & Frey, 2004).

10. As in Eckel and Grossman (2004), donations in the present analysis were not higher among those with various types of religious affiliation than among the nonreligious. Additional analyses including the frequency of church attendance revealed positive effects of being Catholic, Reformed Protestant, or Rereformed Protestant, and a negative effect of church attendance on the probability of making a small donation (1–45%).

11. Additional analyses reveal that males more often made donations than females in the no-subsidy conditions, but not in the experimental conditions in which the price of giving was reduced. Thus, females were more sensitive to the price of giving than males. This result is at odds with the evidence from a lab experiment reported by Andreoni and Vesterlund (2001), who found that males increased their giving more as the price of giving was reduced, while females did not.

12. In a model including expectations but not interactions of expectations with subsidies, expectations had no effect, and the effect of the 33% rebate was still significant. This shows that expectations do not mediate the effect of subsidies on subsequent giving.

13. Another reason why matches are more effective than rebates may be the expectation of transaction costs. In ordinary life, rebates often involve filling out forms to get your money back. Or donors may doubt that they will actually get the

rebate. This possibility is ruled out by the experimental design. Donors did not have to do anything in order to get the rebate.

## ACKNOWLEDGMENTS

This research was supported by a "Veni"-grant (#451–04–110) from the Netherlands Organization for Scientific Research (NWO) and the Royal Academy of Arts and Sciences/Van der Gaag Stichting. The data collection for this study was funded by grants from the Ministry of Justice, the Ministry of Education, Culture and Science, the Ministry of Housing, Spatial Planning and Environmental Affairs, the Ministry of Agriculture, Nature and Food Quality, and the Centre for Global Citizenship (NCDO) to the Center for Philanthropic Studies and matching grants from TNS/NIPO, which conducted the fieldwork. Parts of the data presented here were presented in Bekkers (2006). A previous version of the current paper was presented at the 33rd Annual Arnova Conference, Washington DC, November 17–20, 2005. I thank the editor, an anonymous reviewer, Stephan Meier, and Dave Verkaik for their helpful comments.

## REFERENCES

Alwin, D. F. (1991). Family of origin and cohort differences in verbal ability. *American Sociological Review*, *56*, 625–638.

Andreoni, J., & Vesterlund, L. (2001). Which is the fair sex? On male-female differences in altruism. *Quarterly Journal of Economics*, *116*, 293–312.

Bekkers, R. (2003). Trust, accreditation, and philanthropy in the Netherlands. *Nonprofit & Voluntary Sector Quarterly*, *32*, 596–615.

Bekkers, R. (2006). Effectiviteit van subsidies voor giften aan goede doelen. *ESB*, *91*(4477), 8–10.

Bekkers, R. (2007). Measuring altruistic behavior in surveys: The all-or-nothing dictator game. *Survey Research Methods*, *1*(3), 139–144.

Bekkers, R. (2008). Straight from the heart. In S. Chambré & M. Goldner (Eds.), *Advances in medical sociology: Patients, consumers and civil society: US and international perspectives* (Vol. 10, pp. 197–221). Advances in Medical Sociology. Bingley, UK: Emerald Group Publishing.

Bekkers, R., Boonstoppel, E., & De Wit, A. (2013). *Giving in the Netherlands panel survey – user manual, version 2.2*. Amsterdam: Center for Philanthropic Studies, VU University.

Bekkers, R., & Schuyt, T. N. M. (2008). 'And who is your neighbor?' Explaining the effect of religion on charitable giving and volunteering. *Review of Religious Research, 50*(1), 74–96.

Bekkers, R., & Wiepking, P. (2011). Who gives? A literature review of predictors of charitable giving. part one: Religion, education, age, and socialisation. *Voluntary Sector Review, 2*(3), 337–365.

Carlsson, F., He, H., & Martinsson, P. (2013). Easy come, easy go—the role of windfall money in lab and field experiments. *Experimental Economics, 16*, 190–207.

Central Bureau of Fundraising. (2003). *Verslag fondsenwerving 2002*. Amsterdam: CBF.

Cherry, T. L., Frykblom, P., & Shogren, J. F. (2002). Hardnose the dictator. *American Economic Review, 92*, 1218–1221.

Davis, D. D., Millner, E. L., & Reilly, R. J. (2005). Subsidy schemes and charitable contributions: A closer look. *Experimental Economics, 8*, 85–106.

Deci, E. L., & Ryan, R. M. (1985). *Intrinsic motivation and self-determination in human behavior*. New York, NY: Plenum Press.

Eckel, C. C., & Grossman, P. J. (2003). Rebate versus matching: Does how we subsidize charitable contributions matter? *Journal of Public Economics, 87*, 681–701.

Eckel, C. C., & Grossman, P. J. (2004). Giving to secular causes by the religious and nonreligious: An experimental test of the responsiveness of giving to subsidies. *Nonprofit and Voluntary Sector Quarterly, 33*, 271–289.

Frey, B. S. (1997). *Not just for the money: An economic theory of personal motivation*. Cheltenham: Edward Elgar.

Frey, B. S., & Meier, S. (2004). Social comparisons and pro-social behavior: Testing conditional cooperation in a field experiment. *American Economic Review, 94*, 1717–1722.

Gesthuizen, M., & Kraaykamp, G. (2002). Verbal ability of low-educated people in the Netherlands: the downside of educational expansion. *The Netherlands Journal of Social Sciences, 38*, 191–211.

Heckman, J. (1979). Sample selection bias as specification error. *Econometrica, 47*, 153–1562.

Lindeman, E. (1995). *Participatie in vrijwilligerswerk*. Amsterdam: Thela Thesis.

Meier, S. (2007). Do subsidies increase charitable giving in the long run? Matching donations in a field experiment. *Journal of the European Economic Association, 5*, 1203–1222.

Meier, S., & Frey, B. S. (2004). *Matching donations: Subsidizing charitable giving in a field experiment*. Zurich IEER Working Paper No. 181. Retrieved from http://ssrn.com/abstract=508862.

Meijer, M.-M., Bekkers, R., & Schuyt, T. N. M. (2005). *Uit een goed hart? De Hartstichting en haar donateurs*. Amsterdam: Department of Philanthropy, Vrije Universiteit.

Parker, A. M., & Fischhoff, B. (2005). Decision-making competence: External validation through an individual-difference approach. *Journal of Behavioral Decision Making, 18*, 1–27.

Potters, J., Sefton, M., & Vesterlund, L. (2004). After you—Endogenous sequencing in voluntary contribution games. *Journal of Public Economics, 89*(8), 1399–1419.

Reinstein, D., & Riener, G. (2012). Decomposing desert and tangibility effects in a charitable giving experiment. *Experimental Economics, 15*, 229–240.

Schuyt, T. N. M. (2003). *Geven in Nederland: Giften, legaten, sponsoring en vrijwilligerswerk*. Houten: Bohn Stafleu Van Loghum.

Shang, J., & Croson, R. (2009). Field experiments in charitable contribution: The impact of social influence on the voluntary provision of public goods. *Economic Journal, 119*, 1422–1439.

Smith, V. H., Kehoe, M. R., & Cremer, M. E. (1995). The private provision of public goods: Altruism and voluntary giving. *Journal of Public Economics*, *58*, 107–126.

Stanovich, K. E., & West, R. F. (1998). Individual differences in rational thought. *Journal of Experimental Psychology: General*, *127*, 161–188.

Stanovich, K. E., & West, R. F. (2000). Individual differences in reasoning: Implications for the rationality debate. *Behavioral & Brain Sciences*, *23*, 645–726.

Wiepking, P., & Bekkers, R. (2012). Who gives? A literature review of predictors of charitable giving. Part two: Gender, marital status, income, and wealth. *Voluntary Sector Review*, *3*(2), 217–245.

# APPENDIX

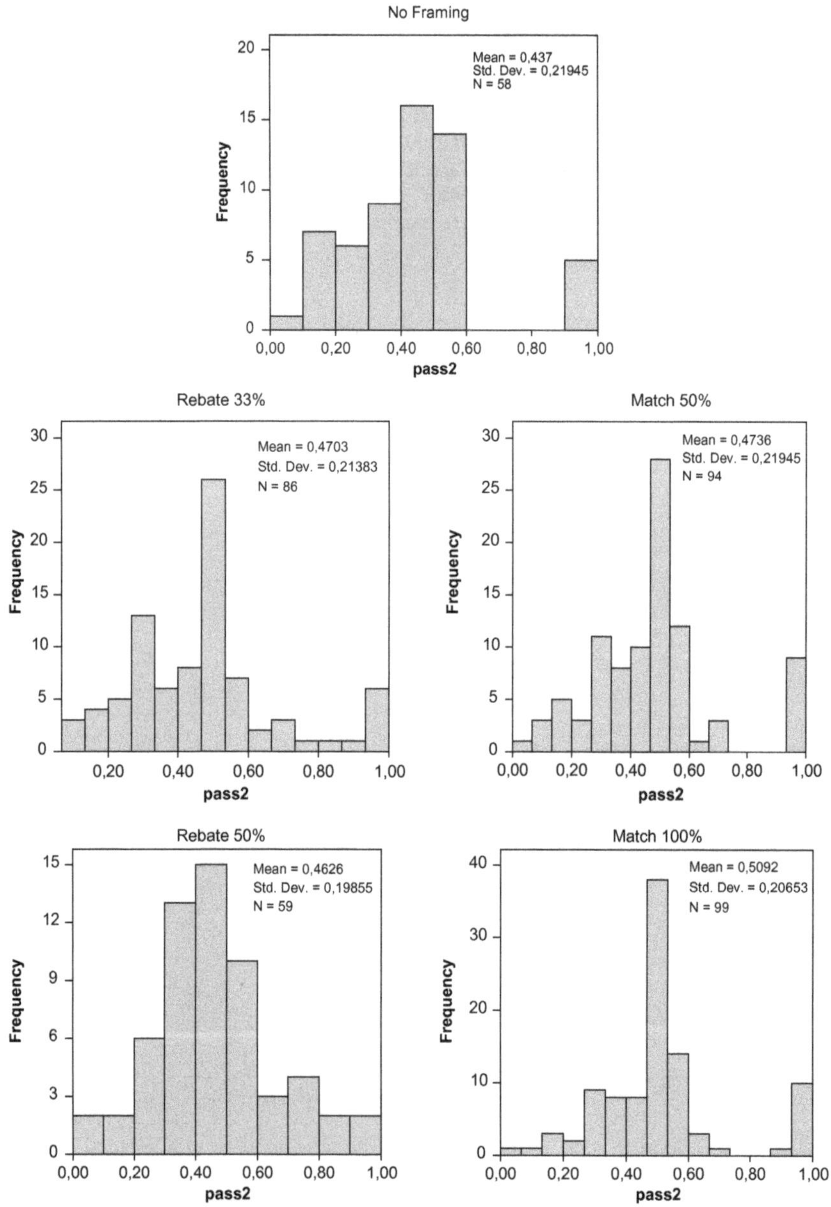


*Figure A1.* Pass Rates in Experiment.

***Table A1.*** Tobit Regressions of Donations on Characteristics of Respondents.

| | No Subsidy | | Match | | Rebate | |
|---|---|---|---|---|---|---|
| | Coeff. | S.E. | Coeff. | S.E. | Coeff. | S.E. |
| Income | −.006 | .003* | .001 | .001 | −.002 | .001 |
| Catholic | .221 | .140 | .095 | .076 | .037 | .090 |
| Reformed | .380 | .163* | .044 | .089 | .125 | .104 |
| Rereformed | .343 | .179(*) | −.056 | .114 | .054 | .131 |
| Other religion | .162 | .261 | .014 | .175 | −.148 | .187 |
| Church attendance | −.006 | .002* | −.001 | .001 | −.001 | .002 |
| Cognitive proficiency | .020 | .024 | .015 | .011 | .032 | .016* |
| Empathic concern | −.257 | .107* | .031 | .058 | −.071 | .068 |
| Joy of giving | .109 | .072 | .064 | .037(*) | .084 | .056 |
| Responsibility | .217 | .088* | .064 | .043 | .091 | .035* |
| Charitable confidence | .106 | .071 | .097 | .040* | .116 | .042** |
| *n* | 165 | | 357 | | 339 | |
| Pseudo *R* Square | .1368 | | .0552 | | .0589 | |

# PART IV
# LOOKING FORWARD

# HOW TO MAKE EXPERIMENTAL ECONOMICS RESEARCH MORE REPRODUCIBLE: LESSONS FROM OTHER DISCIPLINES AND A NEW PROPOSAL

Zacharias Maniadis, Fabio Tufano and John A. List

**ABSTRACT**

*Efforts in the spirit of this special issue aim at improving the reproducibility of experimental economics, in response to the recent discussions regarding the "research reproducibility crisis." We put this endeavor in perspective by summarizing the main ways (to our knowledge) that have been proposed – by researchers from several disciplines – to alleviate the problem. We discuss the scope for economic theory to contribute to evaluating the proposals. We argue that a potential key impediment to replication is the expectation of negative reactions by the authors of the individual study, and suggest that incentives for having one's work replicated should increase.*

**Keywords:** False-positives; reproducibility; replication

**JEL classifications:** B40; C90

**Replication in Experimental Economics**
**Research in Experimental Economics, Volume 18, 215–230**

**ISSN: 0193-2306/doi:10.1108/S0193-230620150000018008**

# INTRODUCTION

Nearly a century ago, Ronald Fisher began laying the groundwork of experimental practice, which remains today. Fisher's work was highlighted by the experimental tripod: the concepts of replication, blocking, and randomization were the foundation on which the analysis of the experiment was based. The current volume of *Research in Experimental Economics* (REE for short) aims at helping experimental economists think more deeply about replication, which has attracted the attention of scholars in other disciplines for some time. In particular, in many biological and human sciences it has been claimed there is a "reproducibility crisis," whereby established results fail to be replicated (Bettis, 2012; Ioannidis, 2005; Jennions & Møller, 2002; Nosek, Spies, & Motyl, 2012).

Even the popular press has joined the discussion, which has in turn attracted the interest of the general public.[1] In contrast, the discussion regarding the importance of replication in experimental economics has only recently been gaining traction (e.g., Coffman & Niederle, 2014; Maniadis, Tufano, & List, 2014). Another initiative in experimental economics is the launch of the new *Journal of the Economic Science Association*, which strives to promote replication studies.

This paper aims to place the endeavor of the current REE volume in context. When it comes to the big picture, the main question is what types of reforms may improve the credibility and replicability of experimental findings. Initiatives like this volume are one of several possible ways in which a culture of frequent replications can be promoted. First, we wish to provide the economics audience with an overview of the solutions and initiatives for the larger problem that have been proposed in comparable disciplines.[2] Second, we discuss the challenges in rigorously evaluating these proposals. Third, we propose a novel solution for inducing replications that uniquely tackles an important issue that is neglected in other approaches to reform: the disincentives of original authors to cooperate with scholars who wish to replicate their studies.

We envision that this analysis will have certain benefits. First, independent scholars as well as interested parties who wish to tackle a possible credibility problem in economics – many of the readers of this volume – will be exposed to several important ideas that are already out there but are difficult to access across disciplinary boundaries. Second, there is a great need for economic theorists to formalize these proposals. As respectable experts within their own "ivory towers," scholars in each discipline are inclined to propose informally solutions that seem promising. At the same time, there is a scarcity of formal and systematic evaluation of those solutions. As John Ioannidis, a leading proponent of meta-research on the replicability problem, notes: "... it is essential that we obtain as much rigorous evidence as possible, including experimental studies, on how these practices perform in real life and whether they match their theoretical benefits [...]. Otherwise, we run the risk that we may end up with worse scientific credibility than in the current system" (Ioannidis, 2012). Finally, we believe that a crucial impediment for a "healthy" culture of frequent replications, namely the disincentive to have one's work replicated, has not been addressed by the current proposals and our suggestion might contribute to a much needed discussion of incentives. We shall first recite the existing proposals for reform and the challenges to evaluate them; then, we describe our modest proposal.

## FOUR CATEGORIES OF PROPOSALS

Proposed solutions can be placed into several different categories. Some of them focus on improving the accuracy of inference that can be made on the basis of received empirical results. Some arguments and initiatives wish to tie researchers' hands in order to prevent excess flexibility: they propose to make pre-registration of studies obligatory. Others mainly concern reforms to raise disclosure standards in reporting empirical research. Finally, there are proposals for increasing the number of replications. After describing the different suggestions, we discuss efforts and difficulties in evaluating their effects in a rigorous scientific framework.

## PRE-REGISTERING STUDIES

A prominent proposed solution is to force researchers to submit a full plan of their design and analysis prior to conducting the experiment.[3] Miguel et al. (2014) note that although clinical researchers in the United States are forced by law since 2007 to submit their studies in a public database and to post a review of the results, only recently have social scientists followed this example. This is a situation where the outcome of the policy for social welfare is unclear. There is a cost in satisfying the requirements of pre-registration, and many people argue that it is likely to stifle creativity and harm exploratory studies. Others argue that what will happen is simply that people will clearly designate exploratory study for what it is, rather than presenting it as hypothesis-driven research (Miguel et al., 2014).

Important issues shall be able to be clarified by means of formal theory. For instance, Henry (2009) introduces a model of strategic disclosure of research results. Interestingly, he finds that mandatory disclosure rules, such as pre-registration, may be either welfare enhancing or reducing depending on specific cases in which information is either withheld or shared. Similarly, Coffman and Niederle (2014) posit that the benefits of pre-registration largely depend on the specificities of the experimental setting: indeed, they suggest that pre-registration will only be fruitful in extreme cases, such as large and expensive field experiments. To generate such theoretical insights, they use the framework by Ioannidis (2005). Their main idea is that pre-registration operates only through reducing the generic research "bias" of each study, but not the selection of positive findings to be reported after multiple testing. Yet, it is clear that theory formulation significantly enhances our capacity of engaging in careful debate. For example, regarding Coffman and Niederle (2014), one could argue that if pre-registration is connected with funding, or otherwise generates a reputation cost in leaving the study unpublished then the model with "research competition" is no longer appropriate. The reason is that as long as there is at least one pre-registration, the results of the given study will necessarily be published, and it is no longer true that only a positive result will be published (as assumed in the version of the model with $k$ competitive research teams).

## RAISING DISCLOSURE STANDARDS

A large set of authors in both the social and the behavioral sciences has argued that the best way to increase the credibility of experimental results is to force researchers to reveal all relevant information that is required in order to make appropriate inference from the experimental studies. In particular, it has been suggested that each experiment must be accompanied by a complete list of important informational pieces that the researcher needs to provide (Fanelli, 2013; Landis et al., 2012; Miguel et al., 2014; Simmons, Nelson, & Simonsohn, 2011). If researchers truthfully complete this list, then appropriate inference from the generated data can be substantially improved.

Many authors have considered this idea. For example, Miguel et al. (2014) note that disclosure standards "… require researchers to report all measures, manipulations and data exclusions, as well as how they arrived at final sample sizes." Simmons et al. (2011) illustrate thoroughly how easy it is to experimentally prove the existence of a false effect, if essential information is absent. Landis et al. (2012) report evidence that many preclinical experiments are poorly reported, and that this is correlated with overstated findings. They also note that many involved parties and stakeholders need to work together in order to achieve increased transparency in preclinical research. Fanelli (2013) proposes to redefine misconduct in such a way that it incorporates mild forms of questionable research practices (QRP), hence promoting transparency.[4]

The idea behind disclosure requirements is that people will honestly follow them. In particular, it has been argued that the enforcement of the disclosure list will transform "sins of omission" into "sins of commission," and hence they will be avoided (Simmons et al., 2011). In other words, it might be true that researchers may exploit the loopholes of the current system by engaging in QRP, but it is also reasonable to expect that otherwise honest people are not likely to engage in pure deception after the regime changes.[5] Despite the plausibility of this argument, it is worth illustrating the difficulty in evaluating this proposal. Given that the set of incentives will change after the reform, it is difficult to predict how things will evolve. Striking down mild transgressions might make people turn to hard forms of misbehavior, and the overall effect remains unclear.

In an example of how economic theory can help, Gall and Maniadis (2015) show that enforcing a transparency requirement may reduce the overall rate of misconduct. The authors examine an environment with limited attention and a restricted number of top journals where researchers are intrinsically motivated to adhere to honesty in performing research, but are also interested in being published well. There are three types of behavior for researchers who wish to publish their results: presenting the results as they are, engaging in "mild" QRP,[6] and engaging in outright deception. The publication game is modeled as a "rat race," in the sense that the more others cheat, the less likely a given researcher is to publish his/her work. The authors find that somewhat counterintuitively, ruling out mild misbehavior (which is the outcome of enforcing disclosure) robustly decreases the overall rate of misconduct.[7]

We believe that this type of theory is necessary in order to structure the arguments and especially for considering general equilibrium effects of policy changes. There is clearly an open and promising area for research here. The challenge is to have the theory guide our empirical research of the likely effect of the regime change to start a virtuous circle between theoretical and empirical investigations in order to derive possible policy implications.

## SHARING DATA AND MATERIALS

It has been argued that sharing data and materials would increase transparency in science (e.g., Miguel et al., 2014); however, the implementation of such an apparently straightforward change has been at times quite complex and has generated a heated transdisciplinary debate. For instance, legitimate questions can be posed about who should set the sharing rules; who should be expected to enact such rules; what type of data/materials should be shared (e.g., raw vs. processed data); who should have access to the data; how third-party access should be granted, and the like (Doshi, Goodman, & Ioannidis, 2013).

In recent years, significant changes have been put in place in economics. For instance, the data availability policies of top journals, such as *The American Economic Review* and *The Review of Economic Studies* currently require that, prior to publication authors of accepted manuscripts must provide data and other relevant materials in order to make them publically available on the journal's repositories. However, such policies are not yet in place for most other scientific outlets in economics.

Along with journals' repositories, new archiving tools have been developed by several stakeholders to facilitate the sharing of data and materials. This is, for instance, the case with the "Open Science Framework," an online initiative by the "Center for Open Science" aimed at archiving data and materials.[8] Moreover, even when those data and materials are available, reproducibility of published results (also called "routine replication" – see Maniadis et al., 2014) may not be easily obtained. In genetics, Ioannidis et al. (2009) demonstrate that despite the public availability of data, the reproducibility of published results by independent scholars – who are experts of the relevant fields – is limited.

In economics, Dewald, Thursby, and Anderson (1986) report on a data availability project implemented by the *Journal of Money, Credit and Baking*: their main finding is that the frequent occurrence of inadvertent errors makes reproducibility close to impossible. Thus, it seems that sharing of data and materials may surely improve transparency, and perhaps attenuate critical obstacles to replication, but it is not the panacea to all the reproducibility problems of scientific research.

## OTHER IDEAS

The three large categories of proposals recited above are not exclusive. For example, Ioannidis (2014) provides a variety of ideas for reform that warrant careful thought and perhaps formalization. An intriguing one is to reduce the return from obtaining grants and occupying positions of power in terms of academic recognition. On the contrary, he envisions a regime where these resources are "penalized," viewed as opportunities that the researcher will have to utilize by increased productivity. On a final positive note, Nosek et al. (2012) are proponents of "paradigmatic-driven" research, whereby the research material is standardized and altered in a very structured way for each new piece of research. It seems to us that experimental economics already achieves this standard relative to other disciplines such as psychology (Hertwig & Ortmann, 2001).

## ENCOURAGING A REPLICATION CULTURE

The practice of raising disclosure standards and of pre-registration target research bias affects reported results. The third possibility is to trust the traditional method for ensuring the truthfulness of research findings: replication. Replication is perhaps the most powerful tool that we have to change prior to about a behavioral relationship between two variables. To see this point, consider the framework in Maniadis et al. (2014), where $n$ represents the number of associations that are being studied in a specific field, $\pi$ is the fraction of these associations that are actually true, $\alpha$ denotes the typical significance level, and $1-\beta$ denotes the typical power of the experimental design. Scholars are typically interested in the Post-Study Probability (PSP) that the research finding is true.

The PSP can be found as follows: of the $n$ examined associations, $\pi n$ associations will be true, and $(1-\pi)n$ will be false. Among the true ones, $(1-\beta)\pi n$ will be declared as true, whereas among the false associations, $\alpha(1-\pi)n$ will be declared true even though they are false (false-positives). The PSP is simply found by dividing the number of true associations which are declared true by the number of all associations declared true.

$$\mathrm{PSP} = \frac{(1-\beta)\pi}{(1-\beta)\pi + \alpha(1-\pi)} \tag{1}$$

Maniadis et al. (2014) use this framework to show the power of replication.[9] To illustrate, we consider their table 5, reproduced here as Table 1. The authors calculate the probability that anywhere from zero ($i=0$ in Table 1) to four investigations ($i=3$ in Table 1) reject the null, given that the relationship is true and given that it is false. They derive the PSP as the fraction of the true associations over all associations for each level of replication. Empirical results reported in Table 1 show that with just a few independent positive replications, the improvement in PSP is considerable, especially in those cases where the original result was a "surprise" or unexpected.

Exact replications not only show us if the initial results are a by-product of QRP, but also ensure that the inherent randomness and context-specificity of human behavior were not the driver of an initial finding. Yet, it is well known that the incentives for performing replications are weak (Ioannidis, 2005; Nosek et al., 2012).

In one of the few models examining incentives for replication, Kiri, Lacetera, and Zirulia (2014) employ a game-theoretic model to demonstrate that a fraction of non-reproducible studies has to be expected in any state of the world. In particular, if in a given field no such studies are identified, their model implies that there must be insufficient replication efforts. They show formally that incentives of would-be replicators play a key role in fostering replication. On the other hand, according to the model, the frequently raised calls for softening the overall incentives to publish do not manage to affect research quality as hypothesized. Much less is known, however, about the exact mechanisms and incentive types for would-be replicators and authors of the initial study.

Despite the relative lack of evidence, there have been numerous proposals and approaches. The first idea with respect to how replication can be enhanced is embodied in this special REE volume and also in the new *Journal of Economic Science Association*: increase the number of outlets that are friendly to replication studies. Nosek et al. (2012), however, argue that this model may not have strong promise, since if replication continues to be considered a low-status endeavor, then these outlets may themselves be characterized as such and researchers might have low interest in publishing in such journals.

An alternative idea is conducting structured sets of replications on the basis of pre-registration (Nosek & Lakens, 2014) which may incentivize replications by guaranteeing publication.[10] Moreover, in psychology large replication projects with the participation of many collaborators have been organized (called "the Many Labs' Replication Project"; Klein et al., 2014).

***Table 1.*** The PSP Estimates as a Function of Prior Probability ($\pi$), Power, and Number of Replications ($i$).

| $\pi$ | Power = 0.80 | | | | Power = 0.50 | | | | Power = 0.20 | | | |
|---|---|---|---|---|---|---|---|---|---|---|---|---|
| | $i = 0$ | $i = 1$ | $i = 2$ | $i = 3$ | $i = 0$ | $i = 1$ | $i = 2$ | $i = 3$ | $i = 0$ | $i = 1$ | $i = 2$ | $i = 3$ |
| | PSP | | | | | | | | | | | |
| 0.01 | **0.02** | **0.1** | **0.47** | 0.91 | **0.02** | **0.1** | **0.45** | 0.89 | **0.02** | **0.07** | **0.22** | 0.54 |
| 0.02 | **0.05** | **0.19** | 0.64 | 0.95 | **0.05** | **0.19** | 0.63 | 0.94 | **0.04** | **0.13** | **0.36** | 0.71 |
| 0.05 | **0.12** | **0.38** | 0.82 | 0.98 | **0.12** | **0.38** | 0.81 | 0.98 | **0.1** | **0.28** | 0.6 | 0.86 |
| 0.1 | **0.22** | 0.56 | 0.91 | 0.99 | **0.22** | 0.56 | 0.9 | 0.99 | **0.2** | **0.45** | 0.76 | 0.93 |
| 0.2 | **0.38** | 0.74 | 0.96 | 1 | **0.38** | 0.74 | 0.95 | 1 | **0.36** | 0.64 | 0.88 | 0.97 |
| 0.35 | 0.57 | 0.86 | 0.98 | 1 | 0.57 | 0.86 | 0.98 | 1 | 0.55 | 0.8 | 0.94 | 0.98 |
| 0.55 | 0.75 | 0.93 | 0.99 | 1 | 0.75 | 0.93 | 0.99 | 1 | 0.73 | 0.9 | 0.97 | 0.99 |

The bold numbers indicate cases where the PSP is less than half.

This provides a paradigm that could also be used for replicating well-known experimental economics results.[11] In fact, in this vein, the "Behavioral Economics Replication Project" is currently undertaking the replication of 18 prominently published studies in the field of "psychology and economics."[12] It has also been proposed that the editors themselves should make direct replication a necessary prerequisite for publication (Begley & Ellis, 2012; Nosek et al., 2012). Nosek et al. (2012) note that a direct replication might be much easier and practical to enforce if the results are very surprising and the data collection is relatively cheap.

## A NEW PROPOSAL: ADDRESSING THE INCENTIVES OF HAVING ONE'S WORK REPLICATED

There are still many barriers to a healthy replication culture. Some are institutional; others relate to the incentive system, and many others are cultural. When it comes to replication the perception is often that a "witch hunt" is taking place. Initial authors, whose studies are under scrutiny and subject to a replication attempt, feel under attack.[13] This is the case also because – both within the scientific community and beyond it – a failed replication often is interpreted as implying either a poorly conducted study or even malfeasance and scientific misconduct from the part of the researchers of the initial study.

However, this should not be the case. As is clear from Eq. (1) above, failed replications should be viewed as part of the mechanics of empirical inference, which is bound to yield a certain number of false-positives (e.g., Coffman & Niederle, 2014; Ioannidis, 2005; Maniadis et al., 2014). On the other hand, replicators' attitude and behavior often may be interpreted as particularly threatening.[14] Thus, original authors face significant costs and little expected returns from their active collaboration with would-be replicators. The dynamics between original authors and would-be replicators – with no surprise – may be at times tough and irksome.[15] This may constitute a further disincentive to promoting a culture of frequent well-conducted replications, or even to cultivate replication as a "social norm," which is followed with the confidence that (most) members of our scientific community will do the same (e.g., Bicchieri, 2006). To foster such a shift and make replication a healthy "social norm" in our scientific community we expect that several measures should be put in place, after positive appropriate evaluations. Several scholars have focused their attention on would-be replicators' incentives (e.g., Nosek et al., 2012) to promote replications and knowledge accumulation. However, this approach may prove insufficient, or simply result in an excessively slow evolution towards a norm of frequent replication (e.g., Cavalli-Sforza & Feldman, 1981).

We focus on a key issue that to our knowledge has not been addressed – namely the strong existing counter-incentives for original authors to have their work replicated. We propose a solution for increasing the frequency of well-conducted and appropriate replications. In our view, the academy is lacking not only incentives for the replicator, but also for the authors of an initial, original, study to accommodate a possible replication. For instance, Ioannidis (2014) notes that sharing the materials necessary for allowing replications to take place is dis-incentivized in the current academic system. In empirical economics, Camfield and Palmer-Jones (2013) lament the lack of incentives to share tacit knowledge necessary for replication, and Andreoli Versbach and Mueller-Langer (2013) document that economics researchers are loathe to sharing their data voluntarily.

Importantly, in this paper we advocate in favor of institutional reforms that ensure a significant expected benefit to original authors from a replication attempt. This might involve novel solutions that require an attitude change not only by authors, but also by influential stakeholders such as professional associations (e.g., the Economic Science Association) funding agencies (such as the European Research Council), academic departments, and the like. For instance, in order to achieve higher expected benefits for the original authors, scientific journals (especially those run under the auspices of professional associations) should always allow for timely commentaries after a replication attempt. This will require very little journal space while allowing some return in terms of publications and citations to the original authors.[16] Moreover, this practice would allow the careful examination of factors leading to the replication results, thus contributing to knowledge accumulation.[17]

In addition, we support the use of the number of replications of one's experimental work (which we might call "replicitations") as a metric for one's research quality. We find this appropriate since it is clear that if one's work is interesting, then it should attract replication efforts. Instead of counting citations alone, replicitations should also be counted and taken into account for funding and promotion purposes, etc.[18] We further argue that "adversarial collaborations" (in which two research teams proposing contrasting hypothesis attempt to resolve the dispute empirically – e.g., Bateman, Kahneman, Munro, Starmer, & Sugden, 2005) or "daisy chains" of replications (in which a given lab attempts to replicate a finding by its "neighbor" lab while being supervised by a member of the replicated lab – e.g., Young, 2012) should be used more often to cast light on "true" empirical regularities.

Reforms along the lines outlined above will likely strike many birds with one stone: first, the practice of replication shall no longer make enemies, but friends. Second, the quality of replications will increase, since the original authors will devote more time and effort to help with the replication. Third, since, as we envision, a replication could be viewed as an act of goodwill, this might induce people to replicate more. In addition, we hypothesize that if there is a shift into a regime of widespread replication, people will no longer be stigmatized if one or two of their findings are not replicated. For us, noise is unavoidable in behavioral research, and more importantly there are also canonical and inevitable forces in scientific practices (outside the realm of fraud) that may tend to bias results (Nosek et al., 2012).

## CONCLUSIONS

In this paper we contribute to the lively discussion concerning the necessary reforms of scientific practices in the face of the "research reproducibility problem." We make three main points. First, increasing the incentives for replication is just one of the four types of reform proposals that researchers in several disciplines have made. Second, our state of knowledge regarding what these proposals entail could greatly benefit from employing the formal tools of economic analysis, and we illustrate this point using distinct examples. This means that economic theorists should play an active role, but they should also make a serious effort to collaborate in multidisciplinary teams and fuel a virtuous circle between theory and empirical evidence. Third, any solution proposed so far has under-appreciated a potential key variable that distorts scientific practice: the enmity among researchers that replication induces. We propose a set of novel solutions to this problem, noting that formal evaluation of these proposals is also necessary in order to fully appreciate the trade-offs involved.

## NOTES

1. For instance, there have been articles in The Guardian (Goldacre, 2011), The Economist (2013, 2014), The New York Times (Zimmer, 2011), and The New Yorker (Lehrer, 2010).

2. In doing so, without claiming exhaustiveness, we intend to also highlight to the interested reader additional sources reporting on relevant scholarly debates and reform proposals advanced. For example, two recent special issues of the psychology journal *Perspectives on Psychological Science* have been devoted to replicability (vol. 7, issue 6, November 2012; and vol. 9, issue 3, May 2014).

3. In a similar spirit, Dufwenberg and Martinsson (2014) argue in favor or changing the submission process. They maintain that it is only the design that should be submitted, but not the results. This is likely to have an equivalent outcome as pre-registration in the sense that it will disconnect the incentive to publish with how the data look like. It also seems to strike both problems (biased methodological choices at the researcher level and publication bias at the macrolevel) simultaneously.

4. A radical approach with respect to transparency comes from new "workflow tools" that allow researchers to analyze their data, while making their results and workflow accessible to others. Poldrack and Poline (2014) discuss how this idea has been applied in genomics research.

5. This notion is also supported by the survey evidence in Fanelli (2009), who shows that pure scientific fraud is much less common relative to mild forms of questionable research practices.

6. A set of survey studies identifies types of questionable research practices that researchers are likely to engage in (Fanelli, 2009; John, Loewenstein, & Prelec, 2012; List, Bailey, Euzent, & Martin, 2001). Nosek et al. (2012) forcefully argue that human cognition works in such ways that it is entirely ordinary behavior that may distort the scientific evidence, for example, by post-hoc hypothesizing and eventually forgetting that the results were unexpected.

7. Of course, the results of Gall and Maniadis (2015) hinge on a few key assumptions, notably the "tournament structure" of the publication game and the higher cost for hard forms of QRP relative to soft forms of QRP. Clearly, it is also important to evaluate the effects of imposing the disclosure requirement under different modeling environments.

8. For details, see https://osf.io/ [Accessed on April 13, 2015].

9. Such Bayesian arguments have a long history in the biomedical sciences. Similar points in the context of these sciences were made in some of the works (Ioannidis, 2005; Moonesinghe, Khoury, & Janssens, 2007; Wacholder, Chanock, Garcia-Closas, El-ghormli, & Rothman, 2004).

10. Nosek and Lakens (2014) introduce the novel endeavor as follows: "This article introduces the first known journal issue in any discipline consisting exclusively of preregistered replication studies. It demonstrates that replications have substantial value, and that incentives can be changed." They note that the advantage of pre-registration lies in guaranteeing publication, and also in addressing a subtle bias of the referees called CARKing: "When the results are known, evaluation of quality is likely influenced by preexisting beliefs. [...] Motivated reasoning makes it easy to generate stories for why results differed from expectations." Pre-registration of the replications addresses this concern by basing the review on the design only.

11. A similar proposal has been made in the field of neuroscience by Wagenmakers and Forstmann (2014), who argue that editors should have an active role in soliciting replication attempts of famous studies with full pre-registration.

12. The project involves scholars from several institutions (i.e., Caltech, the University of Innsbruck, the Stockholm School of Economics, and the National University of Singapore). For further details, see http://sciencepredictionmarkets.com/ [last accessed on April 10, 2015].

13. It seems that there is a lack of maturity in both scientists and the general public about how to respond to replication. Poldrack and Poline (2014) note: "First, it is important to clearly distinguish the lack of reproducibility (even if due to errors) from scientific misconduct. There has been an unfortunate tendency (especially amongst online discussions) to conflate these two issues, which can lead to a toxic atmosphere of mistrust and accusation that undermines any effective scientific discourse."

14. An "etiquette for replication" has been put forward (e.g., Kahneman, 2014) in order to curb this alleged "bullying" by replicators.

15. We can testify this from recent personal experience, being on *both* sides of the fence! (See Maniadis et al., 2014; Simonsohn, Nelson, & Simmons, 2014; Maniadis, Tufano, & List, 2015.)

16. We expect that citations of commentaries along with original and replication articles should be appropriately enforced by authors, reviewers, and editors.

17. A nice example of this is illustrated in the commentaries that appeared in the psychology journal *Social Psychology* (volume 45, 2014, issue 4), which followed the (registered) replication reports published in the previous issue of that same journal.

18. This may be obtained for instance by introducing one or more JEL codes dedicated to replication studies, which will make the meta-database searches much easier and facilitate the use of relevant information by funding agencies, promotion panels, award committees, and the like. An added benefit is that this will also facilitate careful meta-analytic research using the search engines.

## REFERENCES

Andreoli Versbach, P., & Mueller-Langer, F. (2013). *Open access to data: An ideal professed but not practised.* RatSWD Working Papers No. 215. Retrieved March 2015.

Bateman, I., Kahneman, D., Munro, A., Starmer, C., & Sugden, R. (2005). Testing competing models of loss aversion: An adversarial collaboration. *Journal of Public Economics, 89*(8), 1561–1580.

Begley, C. G., & Ellis, L. M. (2012). Drug development: Raise standards for preclinical cancer research. *Nature 483*(7391), 531–533.

Bettis, R. A. (2012). The search for asterisks: Compromised statistical tests and flawed theories. *Strategic Management Journal, 33*(1), 108–113.

Bicchieri, C. (2006). *The grammar of society: The nature and dynamics of social norms.* Cambridge: Cambridge University Press.

Camfield, L., & Palmer-Jones, R. (2013). Three 'Rs' of econometrics: Repetition, reproduction and replication. *Journal of Development Studies, 49*(12), 1607–1614.

Cavalli-Sforza, L. L., & Feldman, M. W. (1981). *Cultural transmission and evolution: A quantitative approach.* Princeton, NJ: Princeton University Press.

Coffman, L., & Niederle, M. (2014). *Pre-analysis plans are not the solution, replications might be*. Retrieved from http://web.stanford.edu/~niederle/Coffman.Niederle.PAP.JEP.October2014.pdf.

Dewald, W. G., Thursby, J. G., & Anderson, R. G. (1986). Replication in empirical economics: The journal of money, credit and banking project. *American Economic Review*, *76*(4), 587–603.

Doshi, P., Goodman, S. N., & Ioannidis, J. P. A. (2013). Raw data from clinical trials: Within reach? *Trends in Pharmacological Sciences*, *34*(12), 645–647.

Dufwenberg, M., & Martinsson, P. (2014). *Keeping researchers honest: The case for sealed-envelope-submissions*. Mimeo.

Fanelli, D. (2009). How many scientists fabricate and falsify research? A systematic review and meta-analysis of survey data. *PloS One*, *4*(5), e5738.

Fanelli, D. (2013). Redefine misconduct as distorted reporting. *Nature*, *494*(7436), 149.

Gall, T., & Maniadis, Z. (2015). *Evaluating solutions to the problem of false positives*. University of Southampton Discussion Paper in Economics and Econometrics 1504.

Goldacre, B. (2011). Researchers don't mean to exaggerate, but lots of things can distort findings. *The Guardian*, August 2011.

Henry, E. (2009). Strategic disclosure of research results: the cost of proving your honesty. *The Economic Journal*, *119*, 1036–1064.

Hertwig, R., & Ortmann, A. (2001). Experimental practices in economics: A methodological challenge for psychologists? *Behavioral and Brain Sciences*, *24*(3), 383–403.

Ioannidis, J. P. A. (2005). Why most published research findings are false. *PLoS Medicine*, *2*(8), e124.

Ioannidis, J. P. A. (2012). Why science is not necessarily self-correcting. *Perspectives on Psychological Science*, *7*(6), 645–654.

Ioannidis, J. P. A. (2014). How to make more published research true. *PLoS Medicine*, *11*(10), e1001747.

Ioannidis, J. P. A., Allison, D. B., Ball, C. A., Coulibaly, I., Cui, X., Culhane, A. C., & Van Noort, V. (2009). Repeatability of published microarray gene expression analyses. *Nature Genetics*, *41*, 149–155.

Jennions, M. D., & Møller, A. P. (2002). Relationships fade with time: A meta-analysis of temporal trends in publication in ecology and evolution. *Proceedings of the Royal Society of London. Series B: Biological Sciences*, *269*(1486), 43–48.

John, L. K., Loewenstein, G., & Prelec, D. (2012). Measuring the prevalence of questionable research practices with incentives for truth telling. *Psychological Science*, *23*(5), 524–532.

Kahneman, D. (2014). A new etiquette for replication. *Social Psychology*, *45*(4), 299–311.

Kiri, B., Lacetera, N., & Zirulia, L. (2014). *Above a swamp: A theory of high & quality scientific production*. Mimeo.

Klein, R. A., Ratliff, K. A., Vianello, M., Adams, R. B., Jr., Bahník, Š., Bernstein, M. J., & Woodzicka, J. A. (2014). Investigating variation in replicability: A "many labs" replication project. *Social Psychology*, *45*(3), 142.

Landis, S. C., Amara, S. G., Asadullah, K., Austin, C. P., Blumenstein, R., Bradley, E. W., & Silberberg, S. D. (2012). A call for transparent reporting to optimize the predictive value of preclinical research. *Nature*, *490*(7419), 187–191.

Lehrer, J. (2010). The truth wears off. *The New Yorker*, *13*, 52.

List, J. A., Bailey, C. D., Euzent, P. J., & Martin, T. L. (2001). Academic economists behaving badly? A survey on three areas of unethical behaviour. *Economic Inquiry*, *39*(1), 162–170.

Maniadis, Z., Tufano, F., & List, J. A. (2014). One swallow doesn't make a summer: New evidence on anchoring effects. *American Economic Review*, *104*(1), 277–290.

Maniadis, Z., Tufano, F., & List, J. A. (2015). *One swallow doesn't make a summer: Reply to Simonsohn, Nelson and Simmons (2014)*. Mimeo.

Miguel, E., Camerer, C., Casey, K., Cohen, J., Esterling, K. M., Gerber, A., & Van der Laan, M. (2014). Promoting transparency in social science research. *Science*, *343*(6166), 30.

Moonesinghe, R., Khoury, M. J., & Janssens, C. J. W. (2007). Most published research findings are false-but a little replication goes a long way. *PLoS Medicine*, *4*(2), 0218–0221.

Nosek, B. A., & Lakens, D. (2014). Registered reports. *Social Psychology*, *45*(3), 137–141.

Nosek, B. A., Spies, J. R., & Motyl, M. (2012). Scientific utopia II. Restructuring incentives and practices to promote truth over publishability. *Perspectives on Psychological Science*, *7*(6), 615–631.

Poldrack, R. A., & Poline, J-B. (2014). The publication and reproducibility challenges of shared data. *Trends in Cognitive Sciences*, *19*(2), 59–61.

Simmons, J. P., Nelson, L. D., & Simonsohn, U. (2011). False-positive psychology undisclosed flexibility in data collection and analysis allows presenting anything as significant. *Psychological Science*, *22*(11), 1359–1366.

Simonsohn, U., Simmons, J., & Nelson, L. (2014). *Anchoring is not a false-positive: Maniadis, Tufano, and List's (2014) 'failure-to-replicate' is actually entirely consistent with the original*. doi:10.2139/ssrn.2351926

The Economist. (2013). Trouble at the lab. 19 October 2013, pp. 23–27.

The Economist. (2014). Metaphysicians: Sloppy researchers beware. A new institute has you in its sights. March 2014.

Wacholder, S., Chanock, S., Garcia-Closas, M., El-ghormli, L., & Rothman, N. (2004). Assessing the probability that a positive report is false: An approach for molecular epidemiology studies. *Journal of the National Cancer Institute*, *96*(6), 434–442.

Wagenmakers, E.-J., & Forstmann, B. U. (2014). Rewarding high-power replication research. *Cortex*, *51*, 10.

Young, E. (2012). Nobel laureate challenges psychologists to clean up their act. *Nature News*. doi:10.1038/nature.2012.11535

Zimmer, C. (2011). It's science, but not necessarily right. New York Times opinion. *New York Times, New York, NY*.